Cambridge Primary

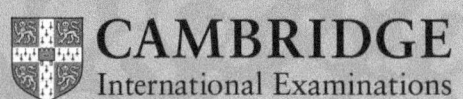

D1823146

Ready to Go Lessons for Science

Step-by-step
lesson plans for
Cambridge Primary

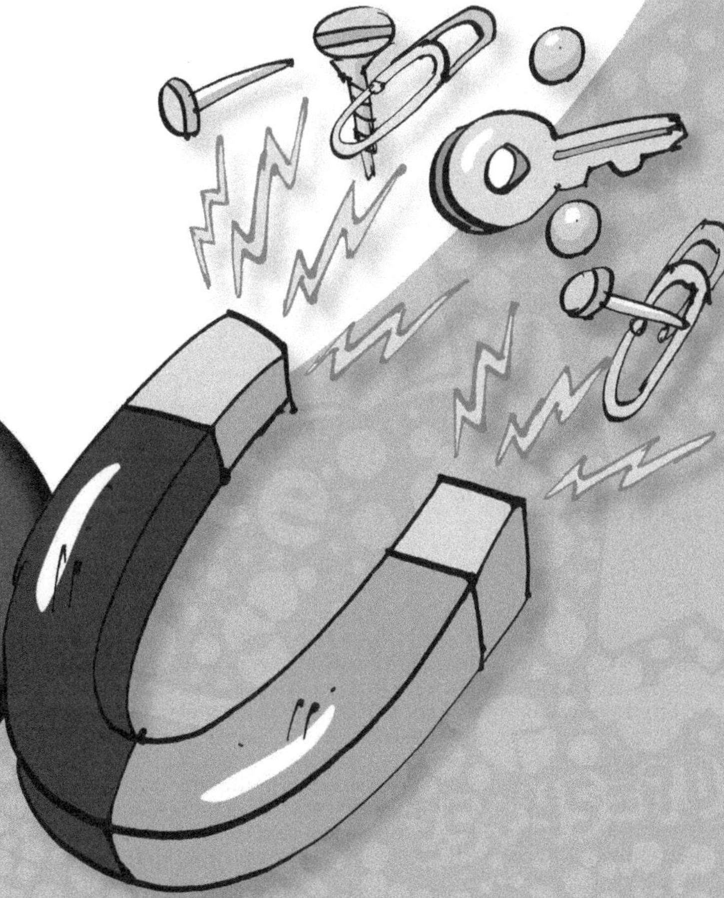

Stage 6

Judith Amery

Series editor: Judith Amery

HODDER
EDUCATION
AN HACHETTE UK COMPANY

Orders: please contact Bookpoint Ltd, 130 Milton Park, Abingdon, Oxon OX14 4SB. Telephone: (44) 01235 827720. Fax: (44) 01235 400454. Lines are open 9.00–5.00, Monday to Saturday, with a 24-hour message answering service. Visit our website at www.hoddereducation.com

© Judith Amery 2013
First published in 2013 by
Hodder Education,
An Hachette UK Company
Carmelite House, 50 Victoria Embankment
London EC4Y 0DZ

Impression number   5 4
Year   2017

Cover illustration by Peter Lubach
Illustrations by Planman Technologies
Typeset in ITC Stone Serif Medium 10/12.5 by Planman Technologies
Printed in Great Britain by CPI Group (UK) Ltd, Croydon, CR0 4YY

A catalogue record for this title is available from the British Library.

ISBN: 978 1444 177879

Contents

Introduction

About the series

Ready to Go Lessons is a series of photocopiable resource books providing creative teaching strategies for primary teachers. These books support the revised Cambridge Primary curriculum frameworks for English, Mathematics and Science at Stages 1–6 (ages 5–11). They have been written by experienced primary teachers to reflect the different teaching approaches recommended in the Cambridge Primary Teacher Guides. The books contain lesson plans and photocopiable support materials, with a wide range of activities and appropriate ideas for assessment and differentiation. As the books are intended for international schools we have taken care to ensure that they are culturally sensitive.

Cambridge Primary

The Cambridge Primary curriculum frameworks show schools how to develop the learners' knowledge, skills and understanding in English, Mathematics and Science. They provide a secure foundation in preparation for the Cambridge Secondary 1 (lower secondary) curriculum. The ideas in this book can also be easily incorporated into existing curriculum frameworks already in your school.

How to use this book

This book covers each of the units of the scheme of work for Science at Stage 6. It can be worked through systematically (as all the learning objectives are covered), or used to support areas where you feel you need more ideas. It is not prescriptive – it gives ideas and suggestions for you to incorporate into your own teaching as you see fit.

Each step-by-step lesson plan shows you the learning objectives you will cover, the resources you will need and how to deliver the lesson.

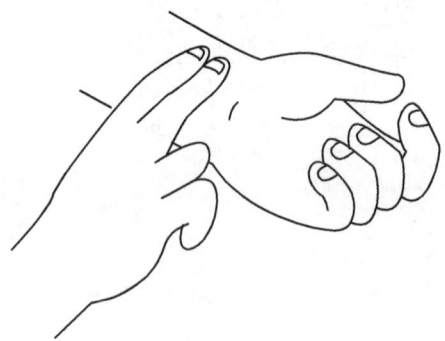

Each lesson includes a Starter activity, Main activities and a Plenary that draws the lesson to a close and recaps the learning objectives. Success criteria are provided in the form of questions to help you assess the learners' level of understanding. The 'Differentiation' section provides support for the less-able learners and extension ideas for the more able.

For each lesson plan there is at least one supporting photocopiable activity page. At the end of each unit there are also suggestions for assessment activities. Answers to activities can be found at www.hoddereducation.com/cambridgeextras.

Learning objectives

The *Science Curriculum Framework* provides a set of learning objectives for each stage. At the start of each lesson you need to re-phrase the learning objectives into child-friendly language so that you can share them with the learners at the outset. It sometimes helps to express them as *We are learning to / about …* statements. This really does help the learners to focus on the lesson's outcomes. For example: 'Know that water is taken in through the roots and transported through the stem' (Stage 3) could be introduced to the learners at the start of the lesson as: *We are learning about the journey water takes through a plant.* To avoid unnecessary repetition we have not included such statements at the start of each lesson plan but it is understood that the teacher would do this.

The overview chart on pages 6–7 shows you how the learning objectives are covered in the lessons in this book.

Time commitment

Teachers should be aware that the recommended time commitment for Science at Stages 1 and 2 is an hour to one and a half hours per week. This could be as a whole afternoon or two or three shorter sessions, depending on timetabling arrangements in your school. The recommended time commitment at Stages 3 to 6 is at least two hours per week. This provides ample time to carry out practical work. Again, it can be timetabled as one long or several shorter sessions. We have, however, provided the same number of lesson plans for you for all six stages to provide choice and variety. Please select the most appropriate lessons for your class for Stage 6 to suit the amount of time available to you.

Success criteria

These are the measures that the teacher and, eventually, the learner will be able to use to assess the outcome of the learning that has taken place in each lesson. They are included as a series of questions, which will help you as teacher to assess the learners' understanding of the skills and knowledge covered in the lesson.

Scientific enquiry skills

Science teaching is concerned with more than just the learning of scientific facts. Scientific enquiry skills are also **essential**.

The activities in these books will show you how to incorporate scientific enquiry skills in order to link practical skills alongside thinking skills using the Cambridge Primary Science Programme. Scientific enquiry is embedded in the curriculum in the Biology, Chemistry and Physics strands. The skills of scientific enquiry are on-going in each stage and between stages. These skills need to be used regularly, in familiar and new contexts, in order for the learners to become young scientists who are capable of questioning, reasoning and finding answers through scientific investigation. Every lesson in this book has links to at least one scientific enquiry learning objective.

The key to successful scientific enquiry teaching lies in providing the learners with opportunities to learn by doing, that is, through **active learning**.

Formative assessment

Formative assessment is on-going assessment that occurs in every lesson and informs the teacher and learners of the progress they are making, linked to the success criteria. The types of questions to ask that will support teachers in making formative assessments have been incorporated into each lesson in the 'Success criteria' sections.

One of the advantages of formative assessment is that any problems that arise during the lesson can be responded to immediately. Formative assessment influences the next steps in learning and may influence changes in planning and / or delivery for subsequent lessons.

Summative assessment

Summative assessment is essential at the end of each unit of work to assess exactly what the learners know, understand and can do. The assessment sections at the end of each unit are designed to provide you with a variety of opportunities to check the learners' understanding of the unit. These activities can include specific questions for teachers to ask, activities for the learners to carry out (independently, in pairs or in groups) or written assessment.

The information gained from both the formative and summative assessment ideas can then be used to inform future planning in order to close any gaps in the learners' understanding as recommended by *Assessment for Learning* (AFL).

Safety

All the lessons in this book have been written with safety in mind. However, please ensure that you are aware of and conform to any national, regional or school regulations for safety as you conduct any of the activities in this book. Always be aware of skin and food allergies or intolerances and obtain parental consent for the learners to participate in tasting activities. If necessary, make sure that you undertake a risk assessment of potential hazards before undertaking activities. It is important to ensure that the learners are aware of safety considerations when carrying out practical activities.

Appropriate use of ICT

At the planning stage teachers need to consider how the use of ICT in a lesson will enhance the learning process. Ensure that the ICT resources you use support and promote the learners' understanding of the learning objectives. Activities included in this book have been designed to be carried out without the need for state-of-the-art ICT facilities. Suggestions have also been included for schools with internet access and / or the use of interactive whiteboards. This is in order to cater for most teachers' needs.

In these lessons the author sometimes asks for the teacher to display an enlarged version of the photocopiable page at the front of the class. We have not specified whether this should be using an overhead projector, interactive whiteboard or flipchart, as schools will have different resources available to them.

We hope that using these resources will give you confidence and creative ideas in delivering the Cambridge Primary curriculum framework.

Judith Amery, Series Editor

Overview chart

The major organs of the human body

- Use scientific names for some major organs of body systems. (6Bh1)
- Make comparisons. (6Eo4)

A model skeleton – small scale or life size; flipchart and markers or whiteboard; internet access or reference books; pencils; photocopiable pages 9 and 10.

Starter

- Bring out the skeleton. Ask the learners to tell you the names of any bones that they can identify. Most of the learners will use general terms, for example leg bone, backbone, ribs, skull.
- Ask if the learners know the scientific name for any bones in the body. They might suggest humerus (funny bone), clavicle (collar bone), patella (knee bone), and so on.
- Ask the learners to discuss with talk partners which parts of our bodies our skeleton protects.
- Listen to their responses and list them as they are mentioned, for example the skull protects the brain, the ribs protect the heart and lungs.
- Invite the learners to share their experiences if any of them have ever had a broken bone, or they know someone who has. How were they treated and how has the broken bone mended? Some of the learners will enjoy sharing the experience of having had an x-ray.

Main activities

- Look at the list you have created and identify that the skull protects the brain and the ribs protect the heart and lungs.
- Ask the learners which other soft parts of the body are **not** protected by the skeleton (stomach and intestines, liver, kidneys).

- Explain that these are called organs and that some of our major organs are: heart, lungs, brain, kidneys, stomach and intestines.
- Show pictures from the internet or books to enable the learners to recognise each organ.
- Give out photocopiable page 9 for all the learners to complete.

Plenary

- Look at the learners' responses to photocopiable page 9 and check that they can identify each organ from its picture.
- Show specific pictures (from the internet or reference books) of different organs, and ask individual learners to name them.
- Emphasise the names of some of the major organs – heart, brain, lungs, kidneys, liver, stomach and intestines.

Ask the learners:

- Which organ does the skull protect?
- Name an organ that is not protected by your skeleton.
- What organ is this / are these? (Point to different organs as you ask this question.)
- Name the major organs in your body.
- Name another bodily organ. (The skin.)
- What do you notice about the organs in another mammal?

Support: Ensure that these learners work in a small group with adult support to complete photocopiable page 9.

Extension: Give these learners photocopiable page 10. This requires them to identify the organs in a small mammal. Ask them to show their completed pages to the rest of the class.

Name: _____

Body organs

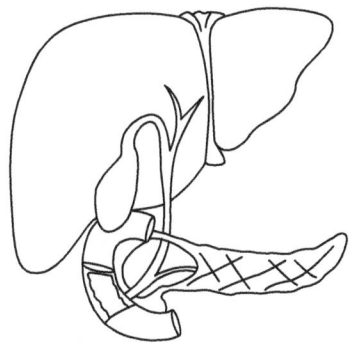

Use these words to name the organs of the body.

| brain | heart | kidneys | liver | lungs |

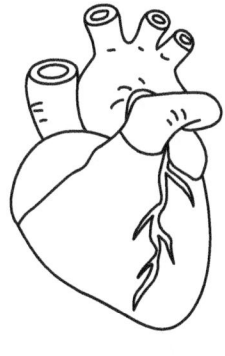

_____

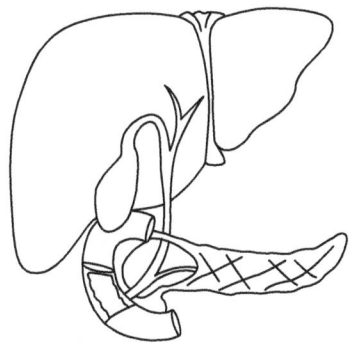

_____

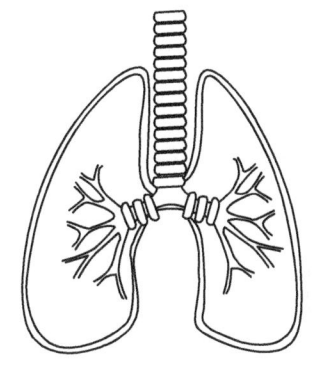

_____

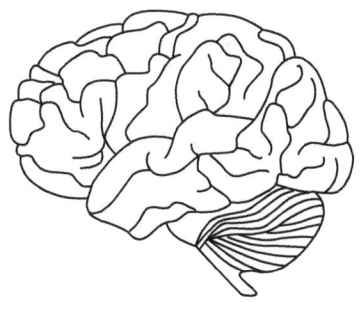

_____

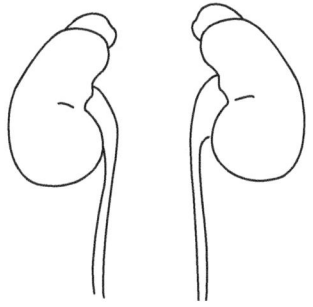

_____

Name: _____

Organs in a small mammal

1. Label the internal organs in this small mammal.

| heart | kidney | liver | lung | stomach |

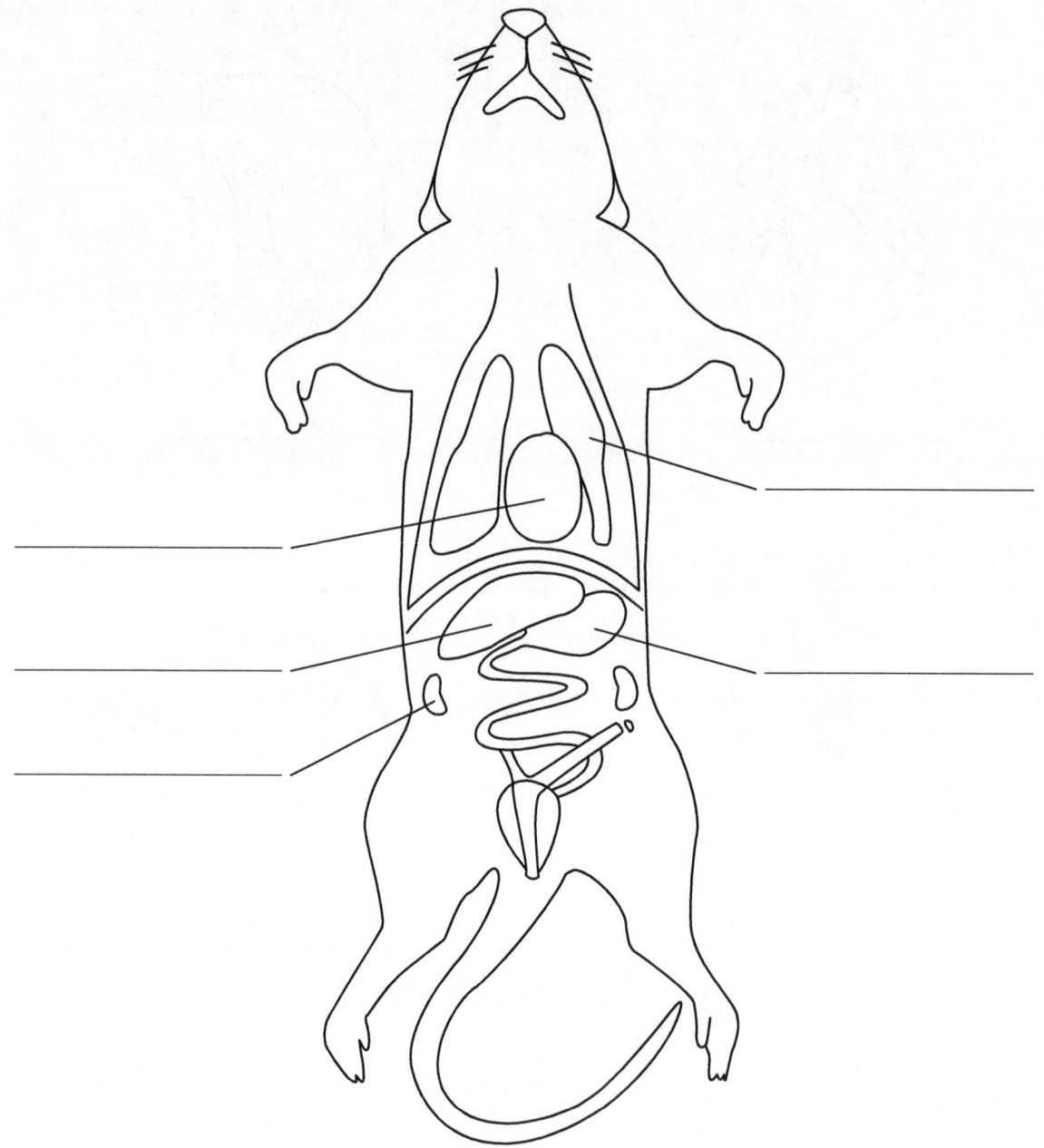

2. How did you know which organs were which?

_____

_____

_____

Cambridge Primary: Ready to Go Lessons for Science Stage 6 © Hodder & Stoughton Ltd 2013

Where are the major organs in our bodies?

● Identify the position of major organs in the body. (6Bh2)

● Make comparisons. (6Eo4)

Photocopiable page 9; reference books or internet access; large pieces of paper; sticky tape; scissors; markers or felt-tipped pens; pencils; coloured paper; 3D fabric organs or 2D paper shapes; a coverall; safety pins or hook-and-loop fasteners (Velcro®); photocopiable page 12.

Starter

• Use photocopiable page 9 from the previous lesson. Name the major organs in the body.

• Show pictures of real organs (from the internet or reference books) and ask the learners to identify them.

• Ask the learners who needed extension in the last lesson to recall how they were able to identify the organs in a small mammal.

• Explain that humans are part of a group of animals called mammals.

Main activities

• Allow the learners in small groups or as a class to draw around the body outline of one learner in their group. They will need several pieces of large paper joined together with sticky tape to do this. Then ask the learners to label or draw where they think the major organs are. Alternatively, they could cut out some paper shapes of organs, predicting their shape and size, and stick these on the body outline.

• Ask each group to show their work and discuss and agree the positions of the organs.

• Alternatively, before the lesson, you could prepare some fabric 3D organs. Attach these to a volunteer learner with safety pins or hook-and-loop fasteners (Velcro®). Ask the learner to wear the coverall and to lie on the floor or a table. Then ask the learners to pretend that they are doctors or surgeons and ask different learners to remove a particular organ.

• Discuss the size, shape and position of the organs in the body.

• Give out photocopiable page 12 and explain to the learners that they have to draw the position of the organs on the body outline.

Plenary

• Look at completed examples of photocopiable page 12 from learners of different abilities.

• Check and comment on their finished pieces of work.

• Choose one body outline from the beginning of the session and either use or re-make 2D shapes to fit and stick onto the outline and display. Alternatively, if there is lots of display space available, each group can adapt its body outline and present it for classroom or corridor display.

Ask the learners:

● Where is your brain?

● Name an organ that you have a pair of.

● Show me with a clenched fist the position of your heart.

● Put your hands on your kidneys.

● On the body outline(s) point to the liver.

Support: Organise these learners to work in mixed-ability groups. Assist them with completing the photocopiable page.

Extension: Ask these learners to use the internet or resource books to find out the positions of the stomach, small intestines and bladder. They could add these to the diagrams on the photocopiable page.

Name: _____

Where are the major organs in our bodies?

Draw and label the following organs in this body outline.

| heart | liver | lungs | kidneys |

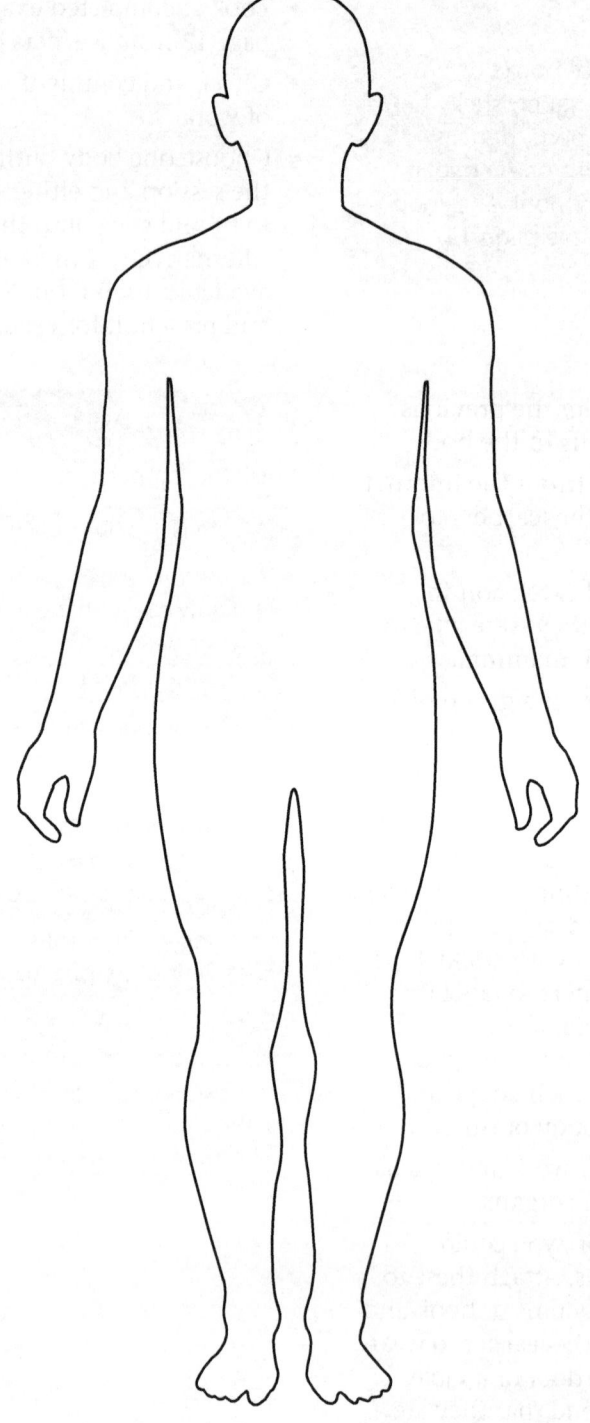

Cambridge Primary: Ready to Go Lessons for Science Stage 6 © Hodder & Stoughton Ltd 2013

The brain and nervous system

Learning objectives

● Describe the main functions of the major organs of the body. (6Bh3)
● Explain how the functions of the major organs are essential. (6Bh4)
● Make comparisons. (6Eo4)

Resources

Resources for using senses (see Starter activity); diagram or poster of the central nervous system; photocopiable pages 14 and 15; internet access or reference books.

Starter

• Ask the learners to discuss with talk partners what their senses are.

• Discuss the responses and identify the five senses – hearing, seeing, taste, touch and smell.

• Prepare some simple activities for the learners to use each of their senses, for example ask them to write down as many sounds as they can hear in a minute or give them some optical illusions to look at. Set up some activities in which they are blindfolded and identify items by taste, touch or smell.

Main activities

• Explain that our sense organs – eyes, skin, nose, ears and tongue – all detect the world around us and help us to make sense of what we see, feel, smell, hear and taste.

• Describe how these senses feed information to our brain. Each sense organ has special cells called sensors. These cells respond to particular stimuli, for example light sensors in our eyes detect light and send a message to the brain along our nerves. Our brain then recognises the stimulus.

• The brain can interpret messages coming to it from all our sense organs. It then sends messages through the nerves to other parts of the body to respond to the stimulus.

• Show the diagram or poster of the main parts of the nervous system.

• Ask the learners to name any parts they recognise, for example the brain, spinal cord, nerves.

• Explain that the brain and spinal cord together are sometimes called the central nervous system, which is sometimes referred to as the CNS.

• Explain that nerves connect to the brain through our spinal cord. If we touch something hot, our brain detects this and sends rapid messages along our nerves for our muscles to contract and pull our hand away from the hot thing. This happens so quickly we are not even aware of thinking about it.

• Give out photocopiable page 14 and ask the learners to answer the questions.

Plenary

• Look again at the poster / diagram and ask the learners to identify the separate parts of the central nervous system.

• Look at pictures (from the internet or books) of the human brain – learners are often fascinated by what it looks like.

Success criteria

Ask the learners:

● What is the main function of our brain? (Control centre for the body.)
● Where is the brain situated in the body?
● How does our brain detect information from around our body?
● What parts make up the central nervous system (CNS)?
● Why is the brain important to us?

Ideas for differentiation

Support: Organise these learners into mixed-ability groups for the Starter activities.

Extension: Challenge these learners to find out the names of different areas of the brain and what they do, for example areas related to memory, smell, and so on. Give these learners photocopiable page 15.

Name: _____

The brain and central nervous system

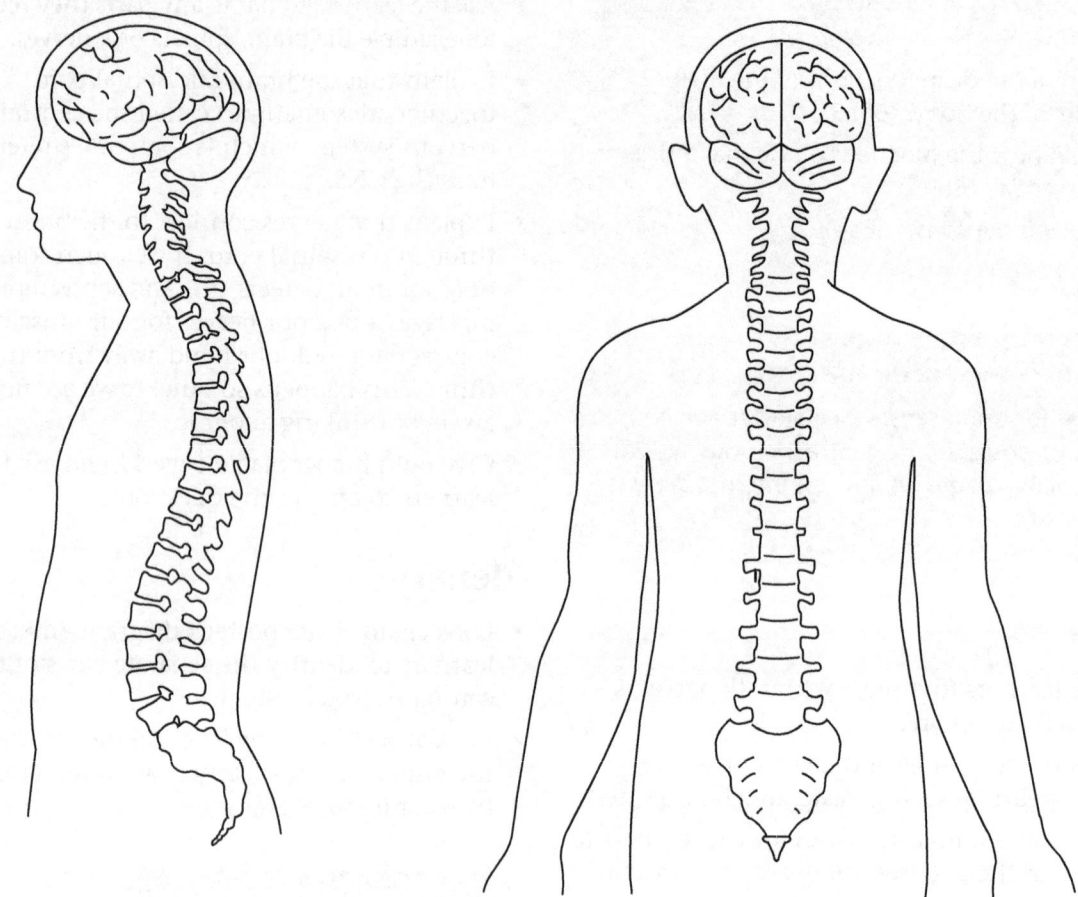

1. What is the main function of the brain in the human body?

    _____

    _____

2. How does information reach the brain?

    _____

    _____

3. How does the brain respond to the instructions it receives?

    _____

4. What makes up the central nervous system (CNS)?

    _____

    _____

Cambridge Primary: Ready to Go Lessons for Science Stage 6 © Hodder & Stoughton Ltd 2013

Name: _____

The brain

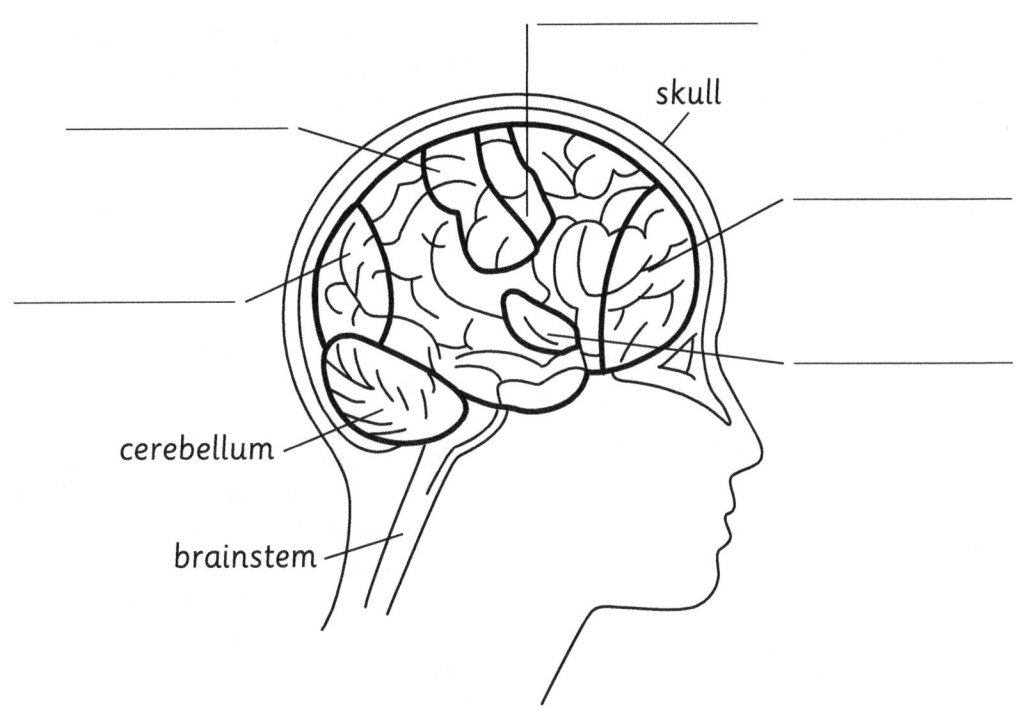

1. Write a few sentences about each of the labelled parts.

   a) Brainstem

   _____

   _____

   _____

   b) Cerebellum

   _____

   _____

   _____

   c) Skull

   _____

   _____

   _____

2. On the diagram, add labels to show which of the senses each part of the brain controls.

How the central nervous system works and why it is important

Learning objectives

● Describe the main functions of the major organs of the body. (6Bh3)

● Explain how the functions of the major organs are essential. (6Bh4)

● Make comparisons. (6Eo4)

Resources

Flipchart and markers or interactive whiteboard; 1.5 kg bag of flour or 1.5 kg jelly in a plastic bag; photocopiable page 17; internet access or reference books.

Starter

• Ask the learners to list with talk partners as many functions (jobs) that the brain does as they can. After a given time limit, share ideas and compile a class list on the flipchart or interactive whiteboard.

• Show the learners the 1.5 kg bag of flour or jelly. Pass it around the class. Tell them that this is how much an adult brain weighs approximately, so if they could handle and compare a real human brain with the bag of flour or jelly, it would be approximately the same weight. Discuss how magnificent it is that such a small part of your body can carry out so many important jobs!

Main activities

• Discuss in detail each suggestion made by the learners in the Starter activity. This could include, for example, cognitive processes (memory / thinking / understanding / imagining), controlling bodily actions, processing signals received via the senses. Explain that the brain is able to collect all the information it receives and process it all the time – even when we are asleep!

• Tell the learners that the brain has two hemispheres. The right hemisphere is responsible for your creative talents, for example if you are good at art, dancing or music, then the right side of your brain is dominant. People with left-hemisphere dominance are usually

good at such things as solving problems, Mathematics and Science. However, we can all do both practical and academic activities, and we are all good at different things.

• Ask the learners to tell you anything that they know that can damage the brain. They will probably mention accidents or injuries, diseases and poisonous substances.

• Think about sports where people wear protective helmets, for example, baseball, horse riding, cycling, and so on.

• Explain that sleep is important for our brain. Ask the learners: *How can we exercise our brain?* (By learning new skills and knowledge and trying to remember and use them.) A healthy diet, rich in fish and vegetables can be particularly good for our brains.

• Ask the learners to complete photocopiable page 17, which considers some things we can do to look after our brain.

Plenary

• Explain that the brain is the control centre for the body, more powerful than any computer yet invented or designed!

• Some injuries and diseases affect a part of the brain and could cause paralysis. (Be aware of any of the learners' family issues, for example if anyone's family member has had a brain injury or disease and is suffering as a result.)

Success criteria

Ask the learners:

● How much does an adult brain weigh?

● How can you protect your brain during sport?

● Name some other ways to keep your brain healthy.

● What is the brain's main function?

Ideas for differentiation

Support: Provide adult support for these learners when completing photocopiable page 17.

Extension: Ask these learners to design a poster showing the brain's structure.

Name: _____

Looking after our brain

1. Draw a meal on the plate showing some foods that are particularly good for our brain.

2. List or draw pictures to show some sports where wearing a helmet is needed to protect the brain.

3. Complete this sentence.

   We can exercise our brain by _____

   _____

4. What does our brain need every day – just like our body?

   _____

The heart and circulatory system

Learning objectives

● Describe the main functions of the major organs of the body. (6Bh3)

● Explain how the functions of the major organs are essential. (6Bh4)

● Make comparisons. (6Eo4)

Resources

Internet access or reference books; photocopiable pages 19 and 20.

Starter

• Give out photocopiable page 19. Ask the learners to mark on the body outline the shape, size and position of the heart.

• Ask the learners to estimate how big their own heart is and to show or tell you and the rest of the class what they think. Invite them to share their responses to photocopiable page 19. This will give you an indication of what they actually know – some of the learners will think that their heart is a perfect heart shape. Many will think that it is in the centre of their body. Do not comment on their responses at this point – simply listen to their ideas.

• Then, tell them to make a clenched fist and explain that that is approximately how big their heart is.

• Now ask them to show you the position of the heart in their body using a clenched fist. Observe their responses and choose a learner who you think has the correct position to come to the front of the class.

Main activities

• Ask this learner to explain their reasoning as to why they think that their heart is in that position. Use this learner as a reference for the rest of the class to observe that the position of the heart is in the middle left of the upper chest.

• Ask the learners to tell you what the main function of the heart is in response to question 2 on photocopiable page 19. (It pumps or sends blood around the body.)

• Explain that as blood travels around the body, it carries in it oxygen and nutrients. It also carries away waste. (This is sufficient detail at this stage, until we go on to look at the excretory and digestive systems later in this unit.)

• Give out photocopiable page 20. Describe the circulatory system, using the diagram on the photocopiable page. Alternatively, show a film clip of the circulatory system to describe this: try www.sciencekids.co.nz/videos/humanbody/circulatorysystem.html. Alternatively, search using your preferred search engine.

Plenary

• Play the film clip again without sound and ask the learners to provide the commentary.

• Discuss and share the learners' responses to photocopiable page 20.

Success criteria

Ask the learners:

● Where is your heart?

● What is the heart's main function?

● What takes blood away from the heart?

● How does blood return to the heart?

● The heart is the major organ in which organ system?

Ideas for differentiation

Support: Discuss these learners' ideas before they complete the Starter photocopiable page. Assist with reading and writing for responses to photocopiable page 20.

Extension: Ask these learners to find out what happens to the heart when it is damaged or diseased and how it can be treated.

Name: _____

Where is my heart?

1.   Draw where you think your heart is on this body outline.

Think about the **shape**, **size** and **position** before drawing it in.

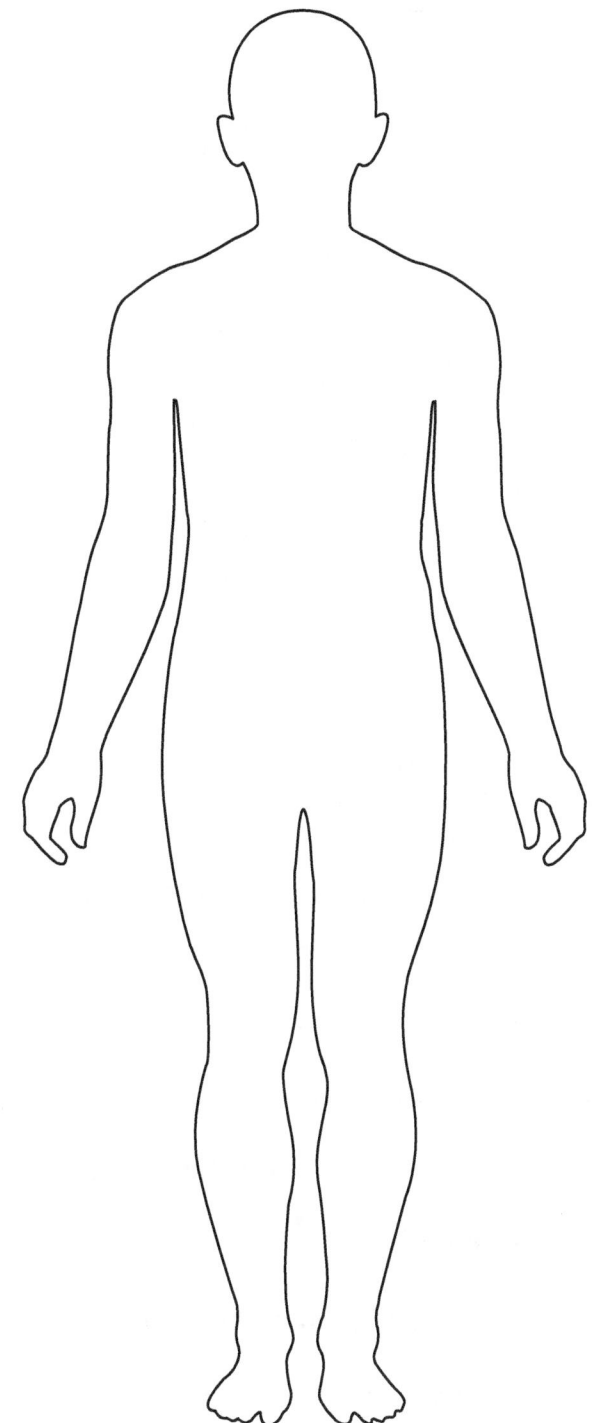

2.   What is the main function of your heart?

_____

_____

Name: _____

The heart and circulatory system

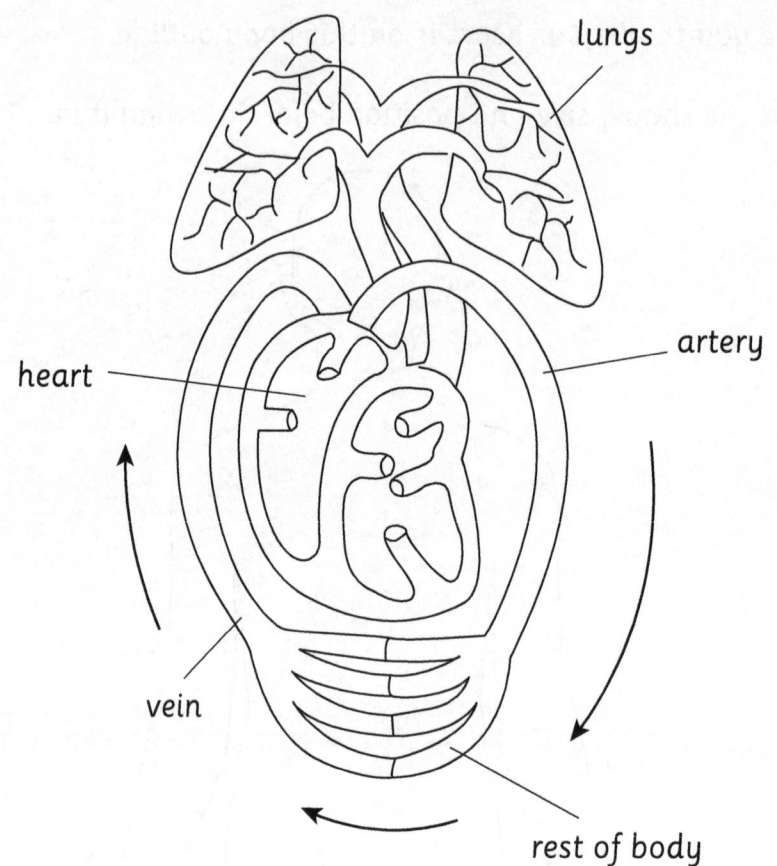

lungs

artery

heart

vein

rest of body

1. Use these words to complete the sentences about the circulatory system. You may need to use some more than once.

| arteries blood body heart lungs muscle pump veins |

a) The heart is a large _____.

b) It acts as a _____.

c) It sends _____ around the _____.

Blood vessels carry blood.

d) _____ carry blood away from the heart.

e) _____ carry blood to the heart.

f) One side of the _____ pumps blood to the _____.

g) The other side of the heart receives blood from the lungs and pumps it

around the _____.

Cambridge Primary: Ready to Go Lessons for Science Stage 6 © Hodder & Stoughton Ltd 2013

Investigating pulse rate

- Describe the main functions of the major organs of the body. (6Bh3)
- Evaluate repeated results. (6Eo5)
- Say if and how evidence supports any predictions made. (6Eo9)

Resources

Photocopiable pages 22 and 23; stopwatches; skipping ropes; music for different dance styles; music player.

Starter

- Ask the learners to discuss with talk partners how they know that their heart is beating.
- Discuss the learners' responses – hopefully, some will mention the pulse rate.
- Describe the pulse rate as a measure of how fast your heart is beating.
- Demonstrate how to feel for or take a pulse in your neck, in the bend of your elbow and in your wrist. (Warning – do not press on the pulse spot too hard!) This can take some time – the wrist may be the best place. Do not feel for the pulse using your thumb (as there is a pulse in it!). Always use two fingers.
- Ensure that the learners are able to find their own pulse easily before going on to the Main activities. They need to be able to count their pulse for a minute or take it for 10 seconds and multiply the answer by 6. Make sure that each learner knows how to use a stopwatch.

Main activities

- Ask the learners: *Which kind of exercise will raise your heart rate the most?*
- Ask them to suggest ideas – they will probably name their favourite sport!
- Now tell them that they are going to investigate which activity raises the heart rate most – skipping, running, hopping, walking or sitting.

- Explain that first they will make a prediction, then plan and carry out a fair test to see if their prediction was correct.
- Discuss what will make this a fair test – factors to keep the same will include such things as the same learner doing the exercise / same time limit each time / taking the pulse rate at the same time (beginning and end) of each activity / same rest period between each exercise. All the activities involved in this investigation can be carried out on the spot.
- Emphasise that the only factor that should change is the type of exercise each time.
- Give out photocopiable pages 22 and 23 for the learners to record their investigation on.

Plenary

- Invite the learners to share their results and discuss if their predictions were correct or not.
- Consider why the results occurred as they did. Explain any misconceptions or misunderstandings.

Success criteria

Ask the learners:

- Which exercise raised your heart rate the most?
- Was your prediction correct?
- Which factors did you keep the same?
- Which factor did you change each time?
- Which factor did you measure?
- What is your conclusion?

Ideas for differentiation

Support: Either work with these learners in a small group or organise them to work in mixed-ability groups.

Extension: Ask these learners to find out how the pulse rate is affected by different styles of dancing.

Name: _____

Investigating pulse rate 1

Prediction

1. (Circle) the exercise that you think will raise your heart beat the most.

hopping   running   skipping   sitting   walking

2. Why do you think this? Give a good, scientific reason.

    _____

    _____

    _____

3. Tick (✓) to show which part of your body you used to take your pulse.

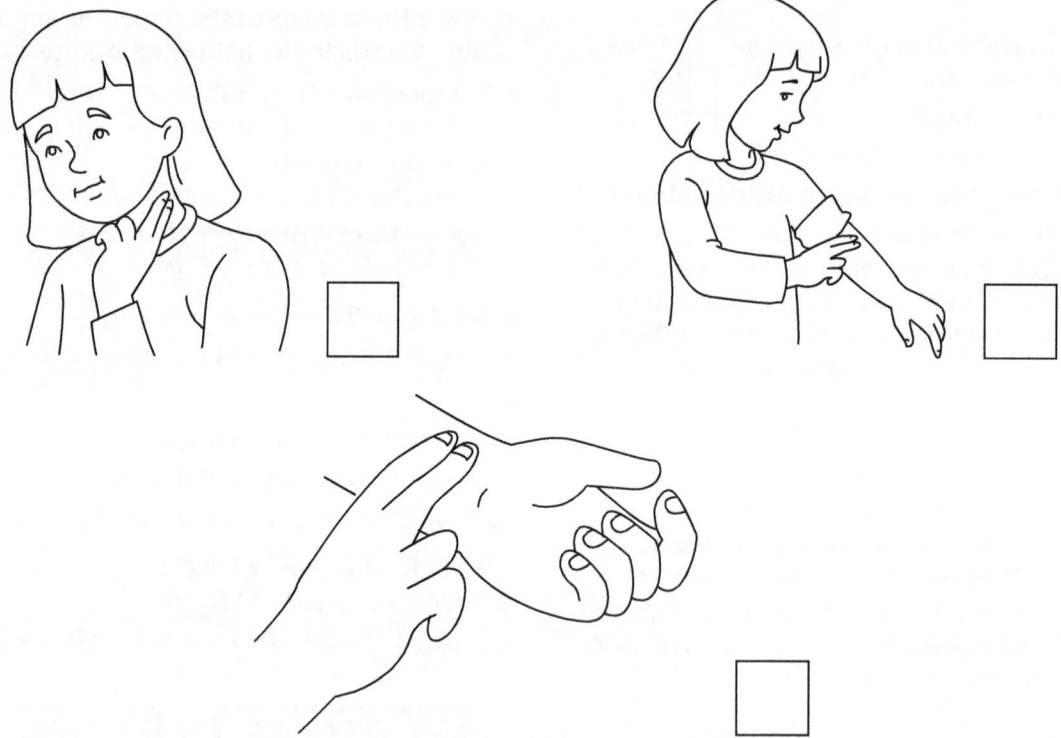

4. What is your pulse rate at rest (when you are sitting still)?

    _____ beats per minute.

5. For how long will you do each exercise? _____

6. When will you take your pulse rate?

    _____

Cambridge Primary: Ready to Go Lessons for Science Stage 6 © Hodder & Stoughton Ltd 2013

Name: _____

Investigating pulse rate 2

Results (what happened)

Pulse rate in beats per minute after exercise				
Hopping	Running	Sitting	Skipping	Walking

Conclusions (what you found out)

1. Which exercise made your heart rate go highest? _____

2. Was your prediction correct? yes / no

3. Do you think that if you did each exercise for longer your pulse rate would be different afterwards? yes / no

   Explain why.

   _____

   _____

   _____

   _____

The lungs and respiratory system

Learning objectives

- Describe the main functions of the major organs of the body. (6Bh3)
- Explain how the functions of the major organs are essential. (6Bh4)
- Make comparisons. (6Eo4)

Resources

Model skeleton or poster; two balloons; drinking straws; sticky tape; internet access or reference books; photocopiable pages 25 and 26.

Starter

- Ask the learners to show on the skeleton or their own body where the lungs are. Ask them to describe what they think the lungs look like.
- Explain that the lungs help us to breathe and that they are like a pair of balloons in our chest cavity. Demonstrate by using a pair of balloons, each with a drinking straw inserted into the neck and stuck with sticky tape. Breathe through the straw to make the balloons inflate and deflate.
- Ask the learners to put their hands on their ribs and breathe in and out deeply. Discuss what they feel and can see when they do this. (The chest raises on breathing in and the ribs push out when we exhale). Do this several times to help their understanding.

Main activities

- Explain that the lungs are part of the respiratory system. They help us to breathe in oxygen and breathe out carbon dioxide. This is sometimes referred to as gaseous exchange – oxygen comes in and carbon dioxide goes out.
- Show a film clip of the respiratory system, for example try http://kidshealth.org/kid/htbw/RSmovie.html. Alternatively, use your own preferred search engine, or describe the process using a model or poster.

- Give out photocopiable pages 25 and 26 for the learners to complete. Photocopiable page 25 requires the learners to label the lungs, see their position within the body and answer some simple questions. Photocopiable page 26 contains more detail about the lungs and the respiratory system.
- Discuss how the lungs are vital for life. If our lungs are not working properly, because they are damaged or diseased, it is sometimes possible to have a lung transplant. This is possible only in certain countries. In these countries, some healthy people carry a special card with them and have their name on a national register to indicate that they want to donate their lungs when they die. Sometimes, a lung donor cannot be found in time and so, sadly, the patient dies from their lung disease. (Be sensitive whenever discussing such issues as some of the learners may not have yet had any experience of death and can easily be upset at the mention of it.)

Plenary

- Watch the film clip again and ask some of the learners to describe what is happening at each stage in the process.

Success criteria

Ask the learners:

- Where are your lungs?
- The lungs are the major organ in which organ system?
- How do your lungs work?
- What scientific process takes place in your lungs? (Gaseous exchange.)

Ideas for differentiation

Support: Provide adult support for these learners when completing the photocopiable pages.

Extension: Ask these learners to find out about any diseases of the lung and how they can be treated.

Name: _____

The lungs

1. Use these words to label the diagram of the lungs.

| bronchial tubes | lungs | trachea (windpipe) |

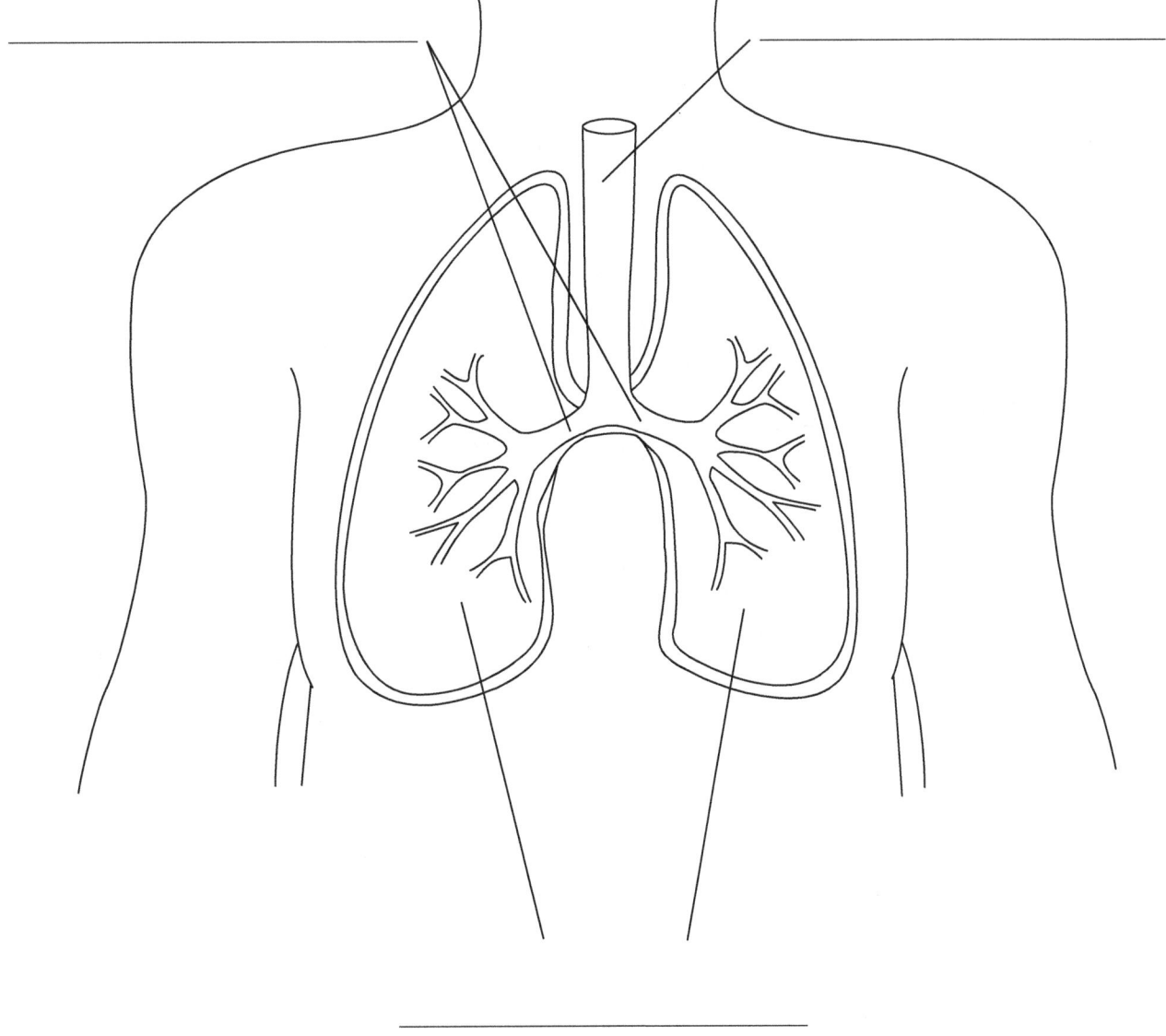

_____

_____

_____

_____

2. What is the gas that we breathe in? _____

3. Name the gas that we breathe out. _____

4. What process takes place inside the lungs? _____

Name: _____

The respiratory system

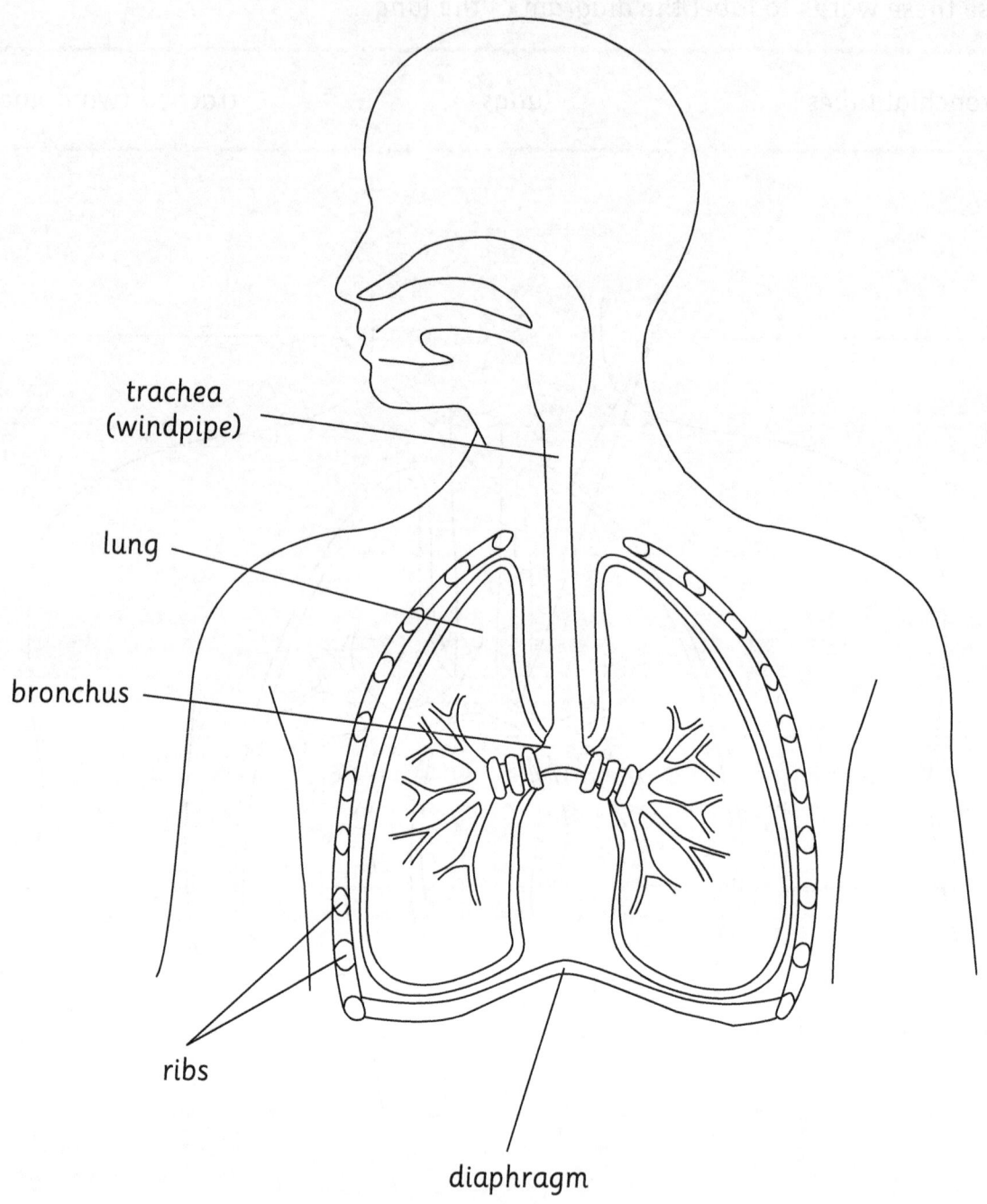

1.   Draw arrows on the diagram to show the journey that oxygen takes when you breathe in.

2.   What route does the gas you breathe out take?

_____

3.   What is the correct scientific term for the process of breathing?

_____

Cambridge Primary: Ready to Go Lessons for Science Stage 6 © Hodder & Stoughton Ltd 2013

The digestive system

Learning objectives

- Describe the main functions of the major organs of the body. (6Bh3)
- Explain how the functions of the major organs are essential. (6Bh4)
- Make comparisons. (6Eo4)

Resources

Sweets or fruit; a plastic bag containing glue and small fruit pieces (tightly sealed); 5 m of broad plastic tubing; 7 m of narrow plastic tubing; photocopiable pages 28 and 29.

Starter

- Give the learners a sweet or small piece of fruit to taste and eat. Ask them to think about the journey that this piece of food will take through their body.

- Think about what is happening in their mouth as they eat – the teeth cut and bite the food, we chew to make it more liquid, saliva softens it and makes it easier to swallow. We taste it! Discuss these things as the learners eat. (Be aware of any food allergies when giving the learners food to eat in class.)

Main activities

- Explain that the digestive system is responsible for digesting our food by breaking it down into the nutrients we need to stay alive.

- Ask the learners to discuss with talk partners the next place our food goes after we swallow it (down the throat to the stomach) and what happens to it there (it gets churned up and stomach acid is added to it to break the food down further).

- Demonstrate the churning action of the stomach by using some thick liquid in a tightly sealed plastic bag with a few solid pieces in it, for example glue with small pieces of fruit in it. Move it around, a bit like washing in a washing machine.

- Then show a film clip of the process of digestion or describe it using other reference materials. Try www.makemegenius.com for free Science film clips and presentations.

- Show the plastic tubing lengths so that the learners can actually see how long the intestines are. Talk about how they are coiled up and tucked neatly inside our abdomen. Challenge the learners who need extension to try to coil both pieces so that they would fit inside an adult body!

- Give out photocopiable page 28 to the learners who need support and photocopiable page 29 to all the other learners, for them to label.

Plenary

- Ask some of the learners to show their labelled diagrams of the digestive system and to explain what happens at each stage.

Success criteria

Ask the learners:

- Where does digestion start?
- Which is the major organ in the digestive system? (The stomach.)
- What happens in the mouth?
- How does the stomach start to digest our food?
- What happens as food travels through the intestines? (Nutrients and water are absorbed into the body.)

Ideas for differentiation

Support: Give these learners photocopiable page 28 to complete.

Extension: Give these learners the task of arranging the plastic tubing into a suitable coiled shape to fit inside an adult.

Name: _____

The digestive system

1. Use the words below to label the diagram.

> food pipe   intestines   mouth   rectum   stomach

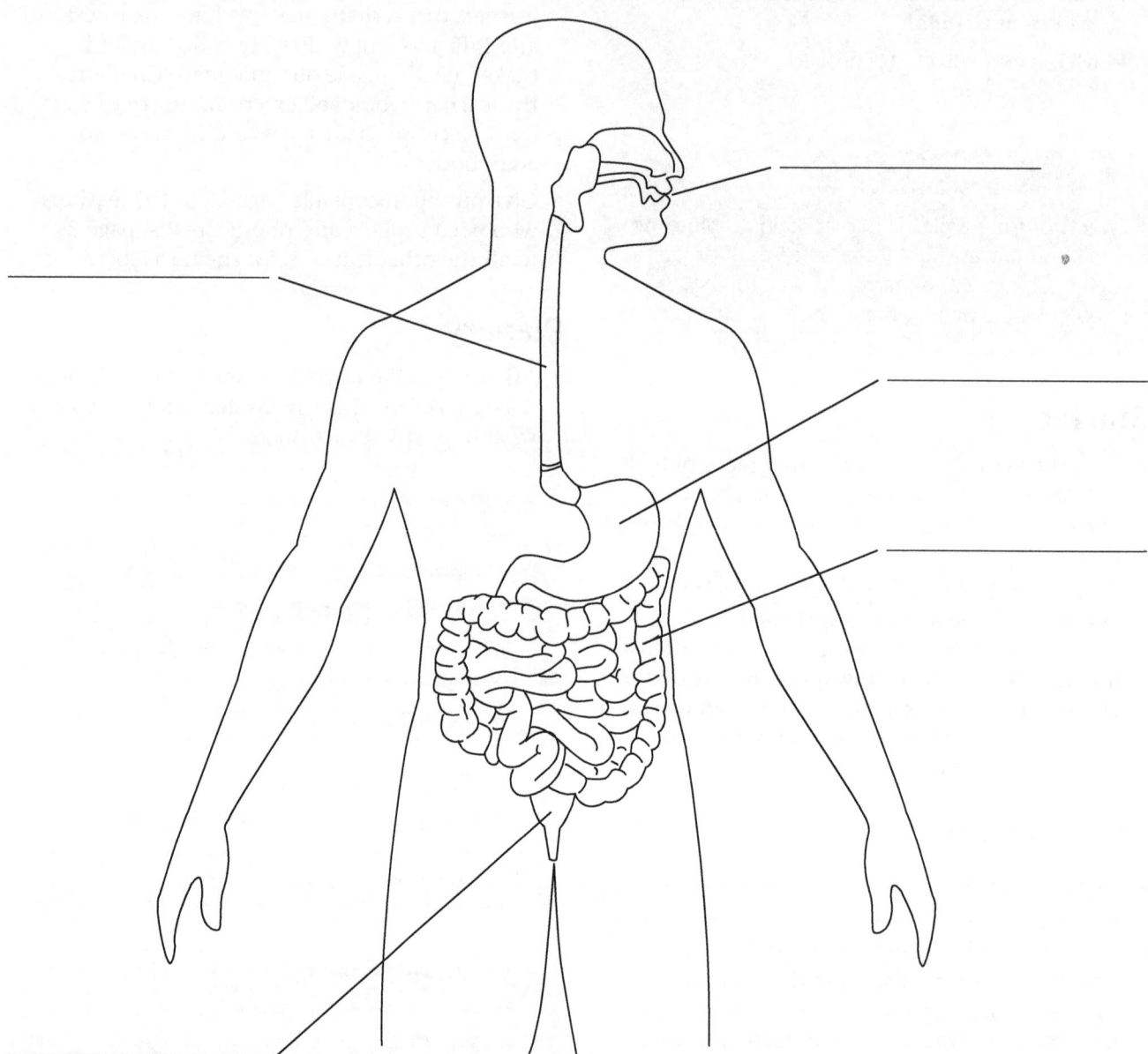

2. What liquid does your mouth produce? _____

3. What happens in the stomach?

_____

_____

Cambridge Primary: Ready to Go Lessons for Science Stage 6 © Hodder & Stoughton Ltd 2013

Name: _____

The digestive system

1. Use the words below to label the diagram.

> anus   food pipe   mouth   large intestine
>
>   liver   rectum   small intestine   stomach

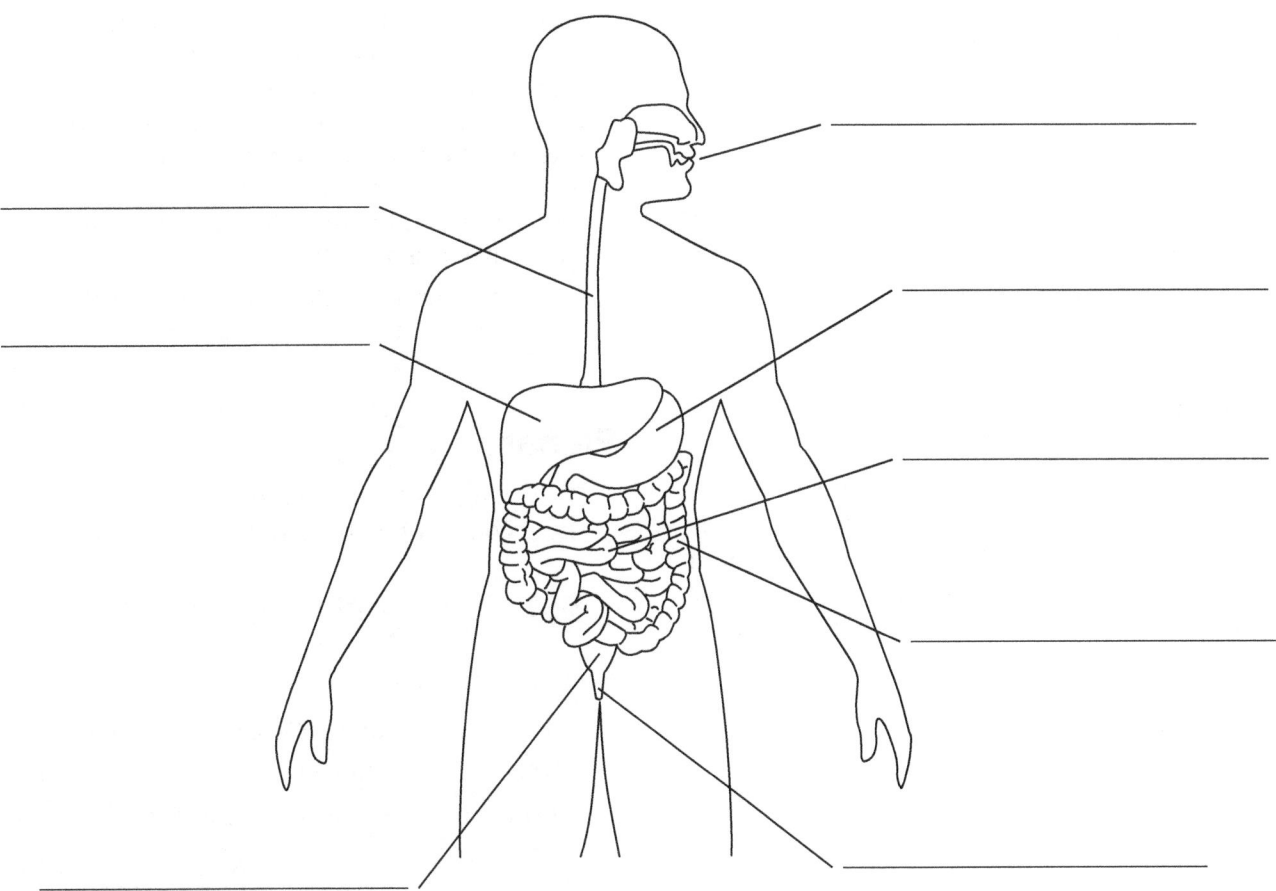

2. Describe the journey of a piece of apple through your body as you digest it. Include each of the words from question 1.

_____

_____

_____

_____

_____

_____

_____

The excretory (urinary) system – the kidneys

Learning objectives

- Describe the main functions of the major organs of the body. (6Bh3)
- Explain how the functions of the major organs are essential. (6Bh4)
- Make comparisons. (6Eo4)

Resources

Body outline(s) produced during the first lesson of this unit; internet access or reference materials; photocopiable pages 31 and 32.

Starter

- Look at the body outline(s) on display produced during the first lesson of this unit. Locate the kidneys. (The kidneys are on both sides of our body near our lower ribs.) They are only about 12 cm long in a fully grown adult. Some of the learners might notice that the kidneys are like kidney beans in shape.

Main activities

- Explain that the liver is an organ that is part of the excretory (urinary) system, which is the body's system for getting rid of waste products. As well as producing bile during digestion, the liver also produces a liquid called urea, which is passed out of our bodies as urine.

- Ask the learners what the function of the kidneys is (they purify or clean our blood – waste products are removed and clean blood returns to the heart). Each kidney contains many tiny filters that filter all the waste the body does not need and send it out of the body as waste.

- Use the internet or reference materials to explain and describe the process of excretion as carried out by the kidneys. Give out photocopiable page 31, which shows the excretory (urinary) system. This is basically the kidneys and the bladder. The bladder stores urine and we get rid of it when we go to the toilet.

- Discuss how you can keep your kidneys in good condition – drink plenty of water, eat a balanced, healthy diet and, as an adult, do not put harmful substances into your body.

- Explain that it is possible to survive with only one kidney. Many people have had successful kidney transplants around the world. Some people who are unable to have kidney transplants regularly use a machine called a dialysis machine, which does the work of filtering their blood as their kidneys are unable to do it properly for them. However, this takes a lot of time and the machines are normally too expensive to have at home.

- Give out photocopiable page 32, which requires the learners to think about ways in which we can look after our kidneys.

Plenary

- Discuss some of the suggestions from the learners when completing photocopiable page 32.

Success criteria

Ask the learners:

- What is the main function of the kidneys?
- Where is urine stored?
- How can a person who has damaged kidneys be helped?
- How can you make sure that you keep your kidneys healthy?

Ideas for differentiation

Support: Discuss these learners' ideas before they complete photocopiable page 32.

Extension: Ask these learners to find out what a dialysis machine looks like and how it works.

Name: _____

The excretory (urinary) system

1. Use the words below to label the parts of this system.

bladder   kidneys   ureter

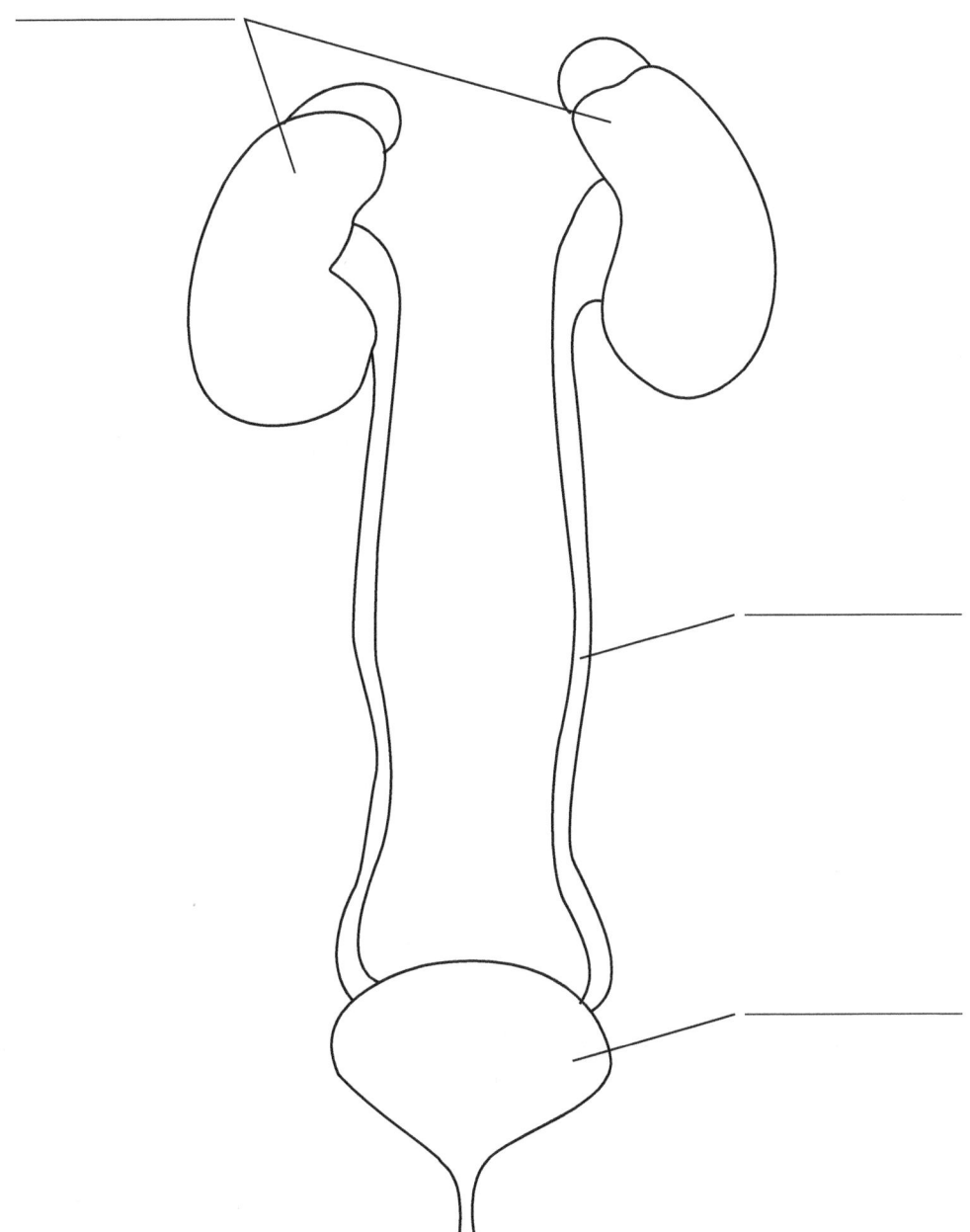

2. What is the main function of the kidneys?

_____

_____

_____

Name: _____

The excretory (urinary) system

Draw pictures and write explanations in the table below about how to keep your kidneys healthy.

Picture Draw something that you need to do to keep your kidneys healthy.	Explanation Why is this good for your kidneys?

 Cambridge Primary: Ready to Go Lessons for Science Stage 6 © Hodder & Stoughton Ltd 2013

Human organs and organ systems

- Describe the main functions of the major organs of the body. (6Bh3)
- Explain how the functions of the major organs are essential. (6Bh4)
- Make comparisons. (6Eo4)

Sticky notes; internet access or reference materials; skeleton or body model; photocopiable pages 34 and 35; a visualiser (if available).

Starter

- Ask the learners to list with talk partners as many major organs as they can in a set time limit, for example 30 seconds. Give them sticky notes to record their responses on.
- Discuss their answers. Who could remember the most? Did they remember them all?

Main activities

- Use the internet or reference materials to find pictures of different organs. Ask the learners to identify them and to explain how they recognise them. Each time a learner gives a response, ask them and the rest of the class to indicate by pointing to the part of their body where that particular organ is situated.
- Alternatively, use the skeleton or body model to do this.
- Next, either use internet film clips or reference materials to revise the functions of the main organ systems; try www.topicbox.net/science/life_body_parts_and_systems/.
  - Use the activity 'Add in the organs' – this asks the learners to place organs inside the body in the order they are given. This is quite a challenging activity and may be useful for the learners who need extension.
  - Try also the 'Animations of the body and body systems and organs'. This includes film clips of the respiratory, nervous, digestive and urinary systems. (It is recommended that you disconnect the sound.)

- If you use only pictorial (book) references, a visualiser can help enable easy viewing of a book for the whole class.
- Allow different learners to attempt the activities, or work through them together as a class. Make sure that you work through these at a pace relevant for the different groups of learners in your class. You might decide to do some as a whole class, for example, and then to work in a small group later in the lesson, repeating all or some of the activities with the learners who need support.
- Give out photocopiable pages 34 and 35. Explain the wordsearch on photocopiable page 34 and the directions in which the words can be found.
- Split the class into pairs or small groups, depending on the availability of resources, to play the game on photocopiable page 35. Read through the instructions with the learners and have a class game to show them how to play it first before allowing them to play their own games.

Plenary

- Revisit some of the interactive activities from the internet site used in the Main activities.

Ask the learners:

- Which is the major organ in the circulatory system?
- The brain is the major organ in which organ system?
- Where are your kidneys situated?
- What is the main function of the digestive system?

Support: Discuss the photocopiable pages with these learners before they complete them.

Extension: Give these learners the 'Add in the organs' activity to do again as a competition. Ask: *Who can complete it in the fastest time?*

Name: _____

Organs and systems wordsearch

Find these words in the grid below.

They can read forwards, backwards, up, down or diagonally.

body   brain   function   heart   kidney   liver   lungs   major   organ

a	s	b	t	c	u	d	v	e	w
h	m	a	j	o	r	x	f	y	g
e	h	z	l	r	a	j	b	k	c
a	l	d	m	g	e	e	f	g	l
r	o	c	s	a	f	v	e	k	u
t	u	f	u	n	c	t	i	o	n
b	r	a	i	n	o	d	e	l	g
k	o	r	t	u	n	l	o	r	s
s	t	d	i	e	m	y	r	s	g
o	l	s	y	s	t	e	m	e	h

Cambridge Primary: Ready to Go Lessons for Science Stage 6 © Hodder & Stoughton Ltd 2013

Organs and systems game

You will need:

A dice, counters, a partner.

How to play

Take turns to roll the dice.

When you land on a square with an organ in it, look at the letter in the same square, and:

- P: point to the **position** in your body of that organ
- S: say which **system** the organ is part of
- F: say the main **function** of that organ.

If you are right, move on three squares. The winner is the first to reach the finish.

64 Finish	63 (brain) S	62	61	60	59 (lungs) S	58	57
49	50	51	52 (kidneys) F	53	54	55	56 (lungs) P
48	47	46	45 (liver) P	44	43	42 (kidneys) S	41
33 (brain) F	34	35	36	37	38 (heart) S	39	40
32	31	30	29 (lungs) F	28	27	26 (liver) F	25
17	18	19	20	21 (kidneys) P	22	23	24
16 (liver) S	15	14	13	12	11 (heart) P	10	9
1 Start	2	3 (brain) P	4	5	6	7	8 (heart) F

Unit assessment

- Name a major organ.
- What is the major organ in the central nervous system?
- The heart is the major organ in which organ system?

- What is the function of the kidneys?
- How can you protect any of your major organs?
- What happens if an organ is diseased or damaged?

Summative assessment activities

Observe the learners while they participate in these activities. You will quickly be able to identify those who appear to be confident and those who may need additional support.

Positions of major organs

This activity assesses the learners' knowledge of the position of the major organs.

You will need:

A skeleton or laboratory body model containing organs; a set of labels naming the major organs.

What to do

- Ask the learners to select an organ or a label and position it in the correct place on the skeleton or model.
- Differentiate this activity by pre-selecting organs for the learners who need support, and challenging the learners who need extension to label as many as they can within a given time.

Functions of organs

This activity assesses the learners' knowledge of the functions of the major organs.

You will need:

Sticky notes; diagrams or models of organs for identification.

What to do

- Ask the learners to choose a picture or model of an organ. Ask them to write on a sticky note a description of the organ's main function, then attach the sticky note to the picture or model in the correct place.

Distribute photocopiable pages 37, 38 and 39 separately (not all at the same time), or select the most appropriate photocopiable page for each learner. The learners should work independently, or with the usual adult support they receive in class.

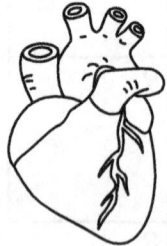
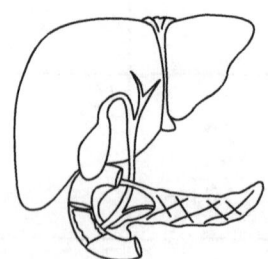
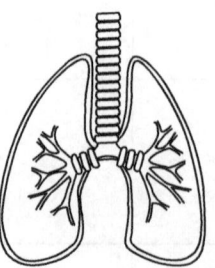
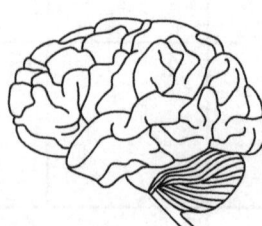
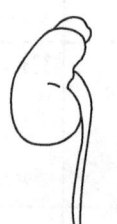
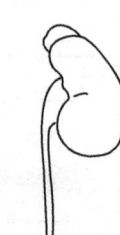

Organs and organ systems

Draw lines to match the organ to the name of the organ system and the function.

One has been done for you.

Organ	Organ system	Function

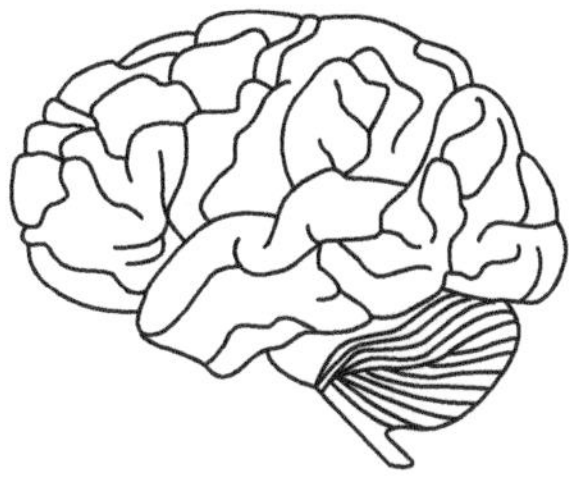

brain

heart

kidneys

stomach

lungs

circulatory

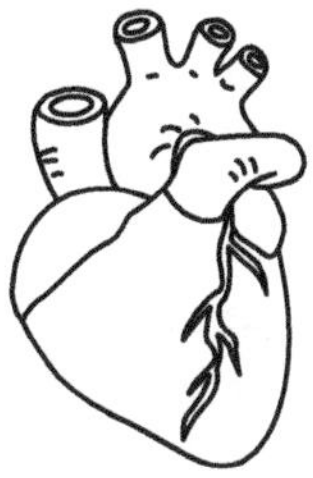

central nervous

respiratory

urinary

digestive

removes waste

pumps blood around the body

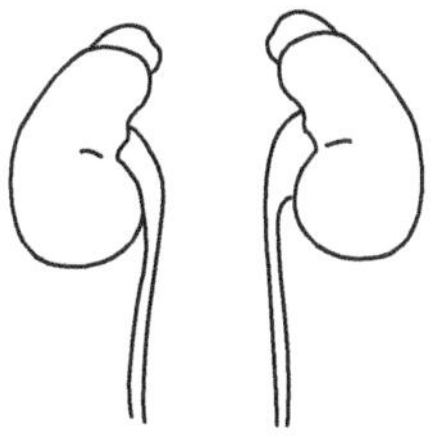

carries messages and acts as the control centre

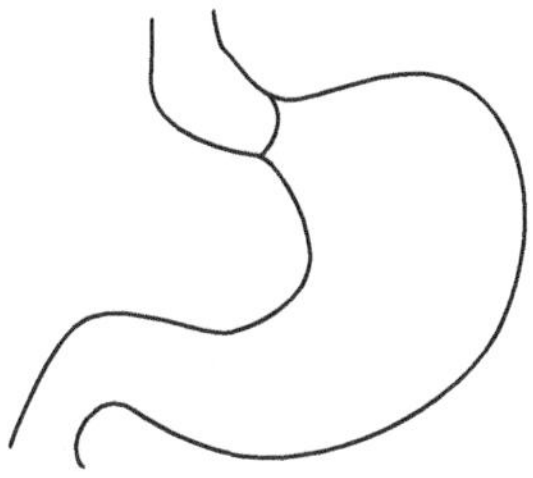

breaks food down

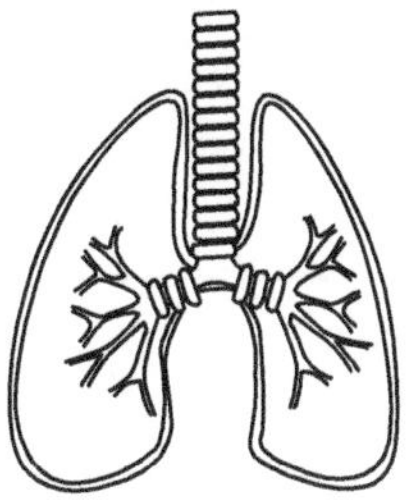

exchange of gases

Name: _____

Where are the major organs?

1. Label the diagram below. Name each of the major organs shown.

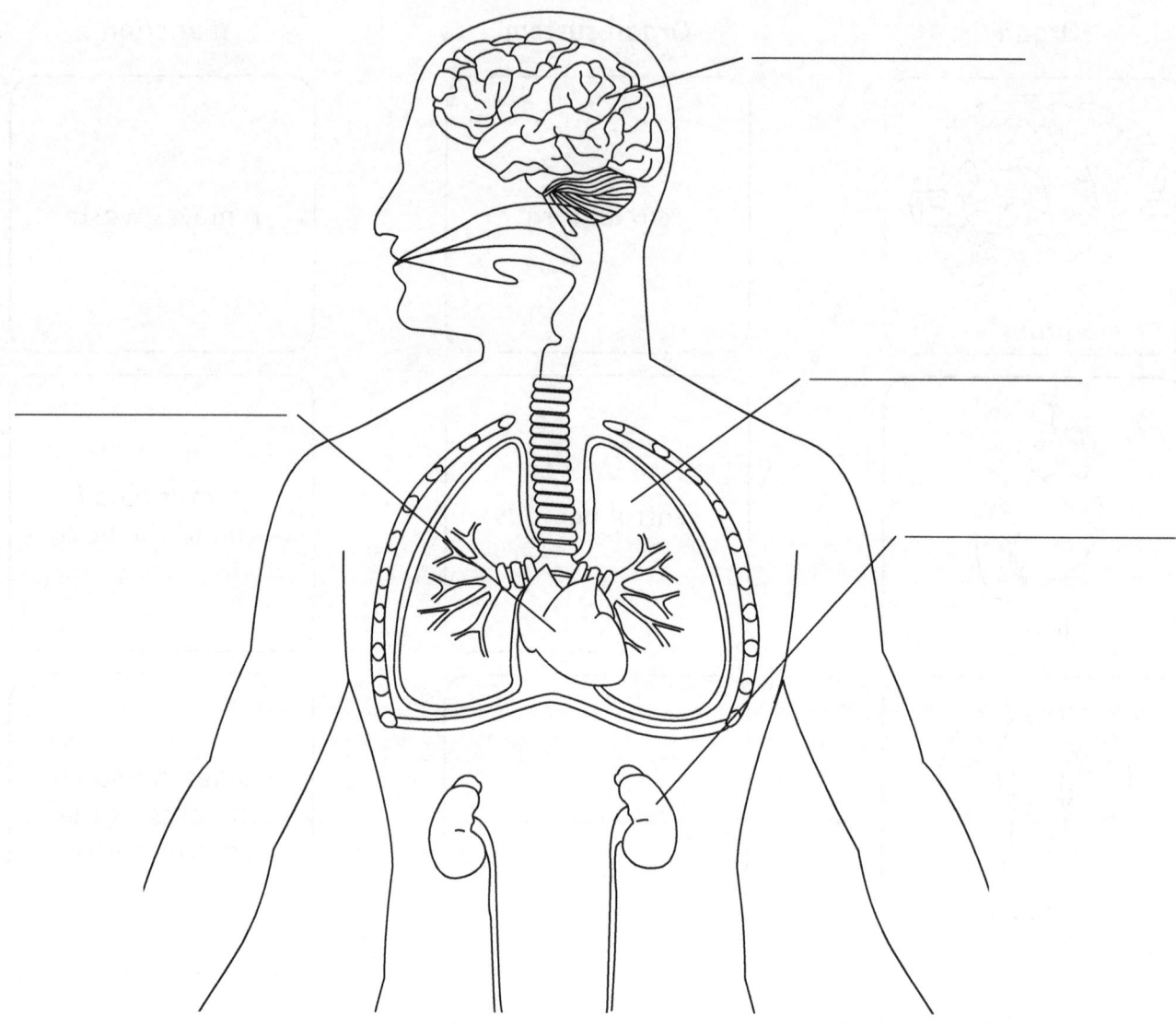

2. Complete the table by naming **two** other organs in the body and their function.

Organ	Function

Cambridge Primary: Ready to Go Lessons for Science Stage 6 © Hodder & Stoughton Ltd 2013

Name: _____

A healthy heart

It is important to keep your organs working well to stay healthy and active.

Design a poster in the box below that tells people how to look after their heart.

Include facts about diet and exercise – as many as you can remember!

Reversible changes

- Distinguish between reversible and irreversible changes. (6Cc1)
- Make predictions using scientific knowledge and understanding. (6Ep4)
- Make comparisons. (6Eo4)

A piece of chocolate for each learner; water and a container; water in a kettle or in a beaker over a heat source for boiling; ice cubes; plastic or metal trays for holding the ice cubes; photocopiable page 41.

Starter

- Introduce this unit as a Chemistry unit in which the learners will conduct lots of experiments to find out about different kinds of changes.

- Give the learners a piece of chocolate to taste (be aware of any food allergies or intolerances). Ask them to predict what will happen to it in their mouth. Tell them not to bite or chew it, but simply to keep it in their mouth. Wait until they have finished eating and then talk about what has happened to the chocolate – it melted. It started off as a solid and melted from the heat of their mouth.

- Ask the learners to think with talk partners of other examples of when chocolate melts, for example in a warm place, in your hand, in the sun or during cooking.

- Remind the learners that the solid chocolate changed to a liquid when it melted.

Main activities

- Revise the concepts of solid, liquid and gas covered in the Stage 4 unit 'Solids, liquids and gases'. Ask the learners to identify things around the classroom as solids, liquids and gases, for example solids – wooden doors, glass windows, plastic rulers; liquids – water, paint; and gases – air, carbon dioxide (from breathing out), and so on.

- Remind the learners that the three states of matter are solid, liquid and gas and that some materials can exist in each state depending on the conditions of the surroundings they are in, especially the temperature of the surroundings.

- Demonstrate or ask a learner to pour some water into a container.

- Ask the learners to discuss with talk partners the different states of matter that water can take.

- Give out photocopiable page 41 and ask the learners to work in pairs to complete it. Give each pair an ice cube to observe. Then demonstrate water boiling, either in a kettle or over a heat source in a beaker. Make sure that the learners are a safe distance away when you boil the water.

Plenary

- Ask the learners to share their responses to photocopiable page 41.

- Explain that water boiling or freezing and chocolate melting are both examples of physical changes that include some kind of change between the different states of matter: solid → liquid → gas or back again, for example gas → liquid → solid.

  Physical changes can be reversed – they are reversible changes.

Ask the learners:

- What does reversible mean?
- What are the three states of matter?
- What happens to chocolate when it gets warm?
- What is the scientific process involved when water turns into water vapour?

Support: Work in a small group with these learners.

Extension: Ask these learners to think about and identify any other everyday changes that involve a change of state.

Name: _____

Reversible changes

Eating chocolate

1. Complete these sentences about eating a piece of chocolate.

    Use these words to help you.

chocolate	liquid	melt	mouth	runny	solid

    a) A piece of _____ is _____.

    b) When you put it in your _____ it begins to _____.

    c) It becomes _____ and turns into a _____.

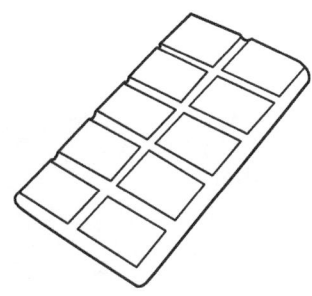

Observing an ice cube

2. Write about the change of state you observe in the ice cube and in the demonstration your teacher shows you. Use the words **solid**, **liquid** and **gas** in your answer.

    _____

    _____

    _____

    _____

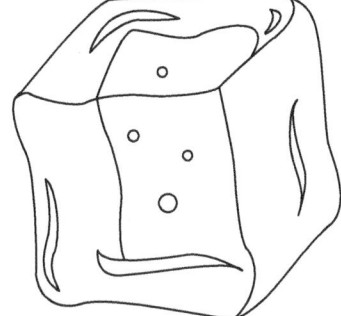

Irreversible changes

- Distinguish between reversible and irreversible changes. (6Cc1)
- Make comparisons. (6Eo4)

Sticky notes; interactive whiteboard or flipchart and markers; photocopiable pages 43, 44 and 45; cake ingredients; access to cooking facilities; cooking aprons; clay; art aprons; clay boards; tools.

Starter

- Ask the learners to write with talk partners a definition of 'reversible'. Give each pair a sticky note on which to write their definition. Either provide a place for the learners to display their definitions when they have written them or invite them to read them out.

- As a class, agree and write a definition of 'reversible change'. Display this prominently for future reference. (Reversible changes are physical changes; physical changes can be reversed.) Solids change to liquids when they melt; they form a solid again on cooling. Liquids form gases when they evaporate, but can be condensed back into liquid again as they cool.

- Explain that in this lesson the learners will find out about different kinds of change – irreversible changes. Discuss what the learners think this might mean. Listen to their responses and tell them that they will be able to check if they were correct in the Plenary session. They might think of cooking and burning as examples.

Main activities

- Give out photocopiable page 43, which gives instructions for making a clay pot.

- Decide if you prefer to split the class into two groups for these activities. The activity on photocopiable page 44 is a baking activity. Half the class could make clay pots whilst the other half do the baking activity. Alternatively, everyone could make a clay pot in this lesson and in the next lesson cover the baking activity.

- Allow the clay-pot making and baking activities to take place. Ensure that the learners take responsibility for tidying up after the activities!

Plenary

- Invite learners from the different groups to talk about what they have made with the rest of the class. If cooking facilities are not available in school, ask the learners to take the ingredients home to cook. They can still talk about the changes that have occurred in preparing the cake mixture.

- Explain that an irreversible change is a permanent change. Irreversible changes are chemical changes. This means that the material changes into something completely new.

Ask the learners:

- What is an irreversible change?
- What does 'chemical change' mean?
- Why is cake an example of an irreversible change?
- Why is making a clay pot an irreversible change?
- Is burning a reversible or irreversible change?

Support: Allow these learners to work in mixed-ability groups for these activities, with adult supervision.

Extension: Give these learners photocopiable page 45 about irreversible changes.

Irreversible changes – making a clay pot

You will need:

An apron, a clay board, a ball of clay.

Choose which type of pot you would like to make.

Coiled clay pot

Method (what to do)

- Knead the clay to get rid of air bubbles.

- Take a lump of clay and roll it into a long roll.

- To make the base start coiling the clay roll.

- Smooth the inside of the pot by pressing clay from the top coil on to the coil below – use your hands to do this.

- Decorate it and leave it to dry or be put in a kiln.

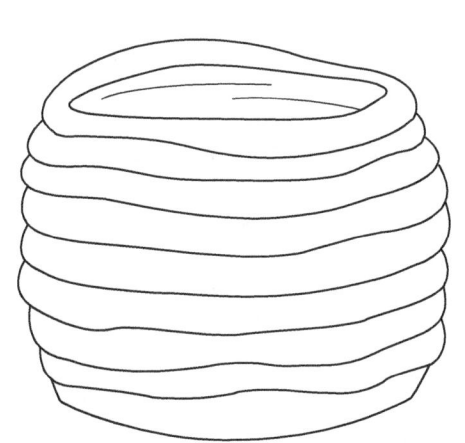

Pinch pot

Method (what to do)

- Take a ball of clay and make it damp.

- Make a hole with your thumb in the middle of the clay.

- Pinch the clay between your finger and thumb and pull upwards.

- Work all the way around the hole.

- Put your hand inside the pot and press firmly onto the table to make the base flat.

- Decorate the outside and leave it to dry or be put in a kiln.

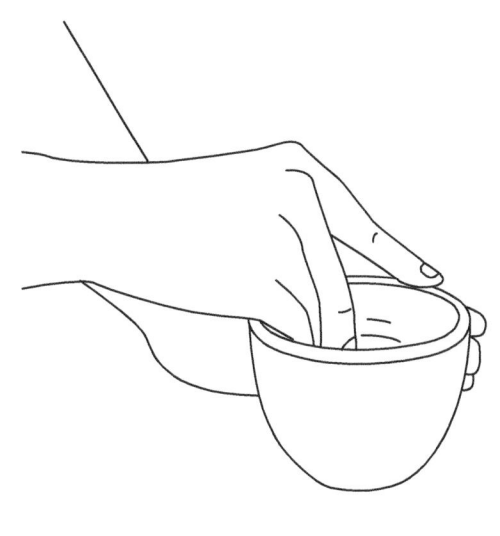

Irreversible changes – baking cakes

Ingredients

For the cakes

200g soft butter

200g caster sugar

4 eggs, beaten

200g self-raising flour

1 teaspoon (5ml) baking powder

2 tablespoons (15ml x 2) milk

For the icing

100g soft butter

140g icing sugar, sifted

a few drops vanilla extract (optional)

Equipment

Baking tins, 12 paper cupcake cases, aprons, wooden spoon or electric whisk, large mixing bowl, cooling rack.

Method (what to do)

1. Heat the oven to 190°C / fan 170°C / gas mark 5. Place the paper cake cases into the baking tins.

2. In a large bowl mix all the cake ingredients until smooth.

3. Spoon the mixture into the cake cases and bake for 15–18 minutes.

4. Cool on a rack.

5. Make the icing by beating the butter until smooth, adding icing sugar and vanilla extract.

6. Decorate the top of each cake with icing.

7. Eat and enjoy!

Cambridge Primary: Ready to Go Lessons for Science Stage 6 © Hodder & Stoughton Ltd 2013

Name: _____

Irreversible changes

Remember! Irreversible changes always make something new.

a candle
burning

a fire
burning

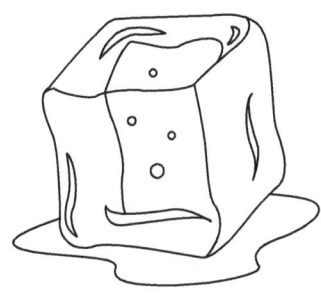

an ice cube
melting

perfume
evaporating

water boiling

1. Look at the pictures.

    Write the **two** irreversible changes.

    a) _____

    b) _____

2. Think about the activity when you baked some cakes.

    Explain why baking cakes is an example of an
    irreversible change.

    _____

    _____

    _____

    _____

Burning – reversible or irreversible change?

Learning objectives

- Distinguish between reversible and irreversible changes. (6Cc1)
- Make comparisons. (6Eo4)

Resources

Clay pots and cakes made previously; safety matches; paper; small sticks; damp sand in a metal tray (for safety); photocopiable page 47.

Starter

- Look at the clay pots made previously. Ask the learners to discuss with talk partners: *Can the clay be returned to its original state*? (No, the clay has hardened.)
- Talk about the cakes made. Ask the learners: *Can you get the original ingredients back*? (No!)
- Explain that cooking and heating are types of chemical change. This is a **permanent** change, which means you cannot get back what you started with – it is irreversible.

Main activities

- Ensure that the learners are seated at a safe distance away to observe these demonstrations.
- Scrunch up a piece of paper to make a ball. Set light to it with a match. Place it in the sand tray until it stops burning. Observe the paper before and after burning. Discuss the changes – it turns into grey ash.
- Make a small fire of sticks in the sand tray. Light the small fire using the safety matches. Again observe the changes during burning – a grey ash is the result.
- Ask the learners if burning is a reversible or irreversible change. Explain that when materials burn they change completely, so burning is another example of an irreversible change.
- Ask the learners to think with talk partners about some examples of when burning something provides us with something that we need (for example burning fuels gives us heat and light).

- Talk about fuels – wood, coal, natural gas, oil. When these fuels burn they can give us light. Petrol or diesel is fuel for our vehicles. Some cars now run on electricity.
- Ask the learners: *What is the fuel for our bodies?* (Food – we **burn** [use up] our food, which gives us energy to move, grow and repair our bodies.)
- Give out photocopiable page 47 and explain that the learners have to think about the before and after products of burning.

Plenary

- Ensure that the learners know that burning is always an irreversible change because when something burns it turns into something completely different.

Success criteria

Ask the learners:

- What is an irreversible change?
- Give an example of an irreversible change.
- Why is burning always an irreversible change?
- What is the benefit to us as humans in burning fuels?
- How do we fuel our bodies and what do we need that fuel for?

Ideas for differentiation

Support: Work with these learners to complete photocopiable page 47.

Extension: Ask these learners what happens when a candle burns. *Is this an example of a reversible or irreversible change?*

Name: _____

Burning

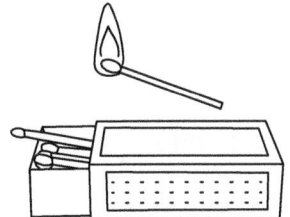

Burning is an example of a **chemical** change.

1. Draw pictures to show what happens when paper and wood burn.

	Before	After
Paper		
Wood		

2. Name **three** natural fuels.

    a) _____

    b) _____

    c) _____

3. How are fuels useful to us?

    _____

    _____

    _____

4. What does your body use as fuel? _____

Burning and melting

- Distinguish between reversible and irreversible changes. (6Cc1)
- Make comparisons. (6Eo4)

Resources

Ice cubes; paper towels; plastic trays or plates; timer; candles; sand trays; safety matches; photocopiable page 49; wax crayons in various colours; sharp pencils.

Starter

- Give each learner (or just a few selected learners) an ice cube to hold. Ask them to keep it solid for as long as possible in their hand. Provide plenty of paper towels for when they want to put it down and something for them to put the melting ice cubes in. Time the ice cube that takes the longest to melt.
- Ask the learners to discuss with talk partners why this ice cube lasted longer than the others.

Main activities

- Explain that heat makes solids melt – even the warmth of your hand on the ice cube and the warmth of your mouth when you eat chocolate or ice cream.
- Ask the learners to tell you again why burning is an irreversible change.
- Tell the learners that they are going to observe a burning candle. Ask them to discuss with talk partners what they need to do to be safe during this activity. Discuss their suggestions and agree exactly what they have to do.
- Give out candles in sand trays to the learners in groups or pairs or individually. Make sure that the candles are secured in holders and placed in sand trays for safety. The candles could be all the same or different so that you can compare the learners' findings at the end of the lesson. Go around and light the candles. On no account should the learners touch the candle once it is lit and ensure that they stand a safe distance away from the candle.

- Give out photocopiable page 49 for the learners to record their observations on.
- Carry out the activity. Circulate the room, checking for safety and asking relevant questions as the learners observe and make notes, for example: *Why is the wax dripping or running down the side of the candle? What can you see in or around the flame?*

Plenary

- Explain that the candle wax melting is a reversible change. Explain that the **burning** candle is an irreversible change. This is a difficult concept for some learners to understand.

Success criteria

Ask the learners:

- What happens when a candle burns?
- Is what happens a reversible or irreversible change?
- What happens to the wax?
- Which part of the candle burns?

Ideas for differentiation

Support: Organise these learners to work in mixed-ability groups, or work with them in a small group.

Extension: Make a wax relief drawing of a burning candle with these learners. Colour the background in wax crayon on a small piece of paper. Cover this entirely with a dark-coloured wax crayon over the top. Using a sharp pencil, draw an image of a burning candle, scraping off the top layer of wax for the outline of the candle and the whole of the flame.

Name: _____

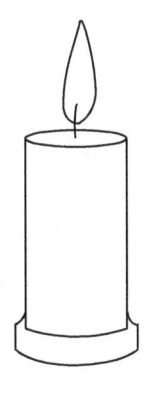

Observing a candle burning

1. Draw a picture of the burning candle.

2. Write a list of words to describe the burning candle around your drawing.

3. Think about what you can see, hear and smell.

4. Look closely at the flame as it burns.

Separating solids

Learning objectives

● Explore how solids can be mixed and how it is often possible to separate them again. (6Cc2)

● Choose which equipment to use. (6Ep7)

● Make a variety of relevant observations and measurements using simple apparatus correctly. (6Eo1)

Resources

Beads of two different colours and maybe sizes; large transparent containers (one for each pair); a variety of sieves of different sizes; magnets; cling film; dried lentils and dried peas, beads, iron filings, rice and sand; photocopiable page 51; salt – but only make this available for the extension activity.

Starter

● Show the learners the beads. In pairs, give one learner beads of one colour and the other learner the other colour. Give them a container to put both sets of beads in. Ask the learners: *What have you made?* (A mixture.)

● Ask the learners to discuss with talk partners if a mixture is something completely different from what they started with (like there was something completely different remaining after the burning experiments).

● Explain that a mixture is just two or more things combined – here there are two different colours of beads.

Main activities

● Tell the learners that in this lesson they will make mixtures and try to separate them.

● Ask the learners: *How could you separate the mixture of beads?* (If they are the same or similar size, they can be separated out by hand. Otherwise, the learners may suggest using a sieve.)

● Show the learners the solids they can use to make their mixtures. Do not discuss how they might separate any mixtures. Either give them prepared mixtures or ask them to make and separate a mixture of their own choice.

● Give out photocopiable page 51 for them to record what they do. Tell them that you need to see the mixture they have made and that they might need to ask you for some equipment to separate the mixture. (It is useful to wrap magnets in cling film so that the substance in the mixture that is attracted comes into contact with the cling film, which can then be removed – this is less difficult than cleaning iron filings off magnets afterwards.)

● Allow the learners to carry out the activity to make and separate a mixture in groups or pairs.

Plenary

● Invite different groups or pairs of learners to share the mixture they made and how they separated it.

● Make sure that the learners realise that all the substances used in this lesson are solids. The best ways to separate mixtures of solids are by sieving, by hand or by using a magnet.

Success criteria

Ask the learners:

● What two things did you put into your mixture?

● Are these things solids, liquids or gases?

● How did you separate them?

● Is this change reversible or irreversible?

Ideas for differentiation

Support: Tell these learners which mixture(s) to make.

Extension: Ask these learners how they could separate a mixture of salt and sand.

Name: _____

Separating solids

Making a mixture

1. We mixed _____ and _____.

2. Write or draw in the box below to show how you separated the mixture.

3. The equipment we used was _____.

4. Write **three** ways you could separate a mixture of solids.

    a) _____

    b) _____

    c) _____

Mixing solids with water

Learning objectives

- Observe, describe, record and begin to explain changes that occur when some solids are added to water. (6Cc3)
- Collect evidence and data to test ideas including predictions. (6Ep2)
- Choose which equipment to use. (6Ep7)

Resources

Salt; sand; beakers or plastic cups; stirrers; teaspoons; sieves; paper towels; filter funnels or cut-off plastic bottles; photocopiable pages 53 and 54.

Starter

- Ask the learners who did the extension activity in the previous lesson how they separated (or would separate) a mixture of salt and sand. If the extension activity was not attempted in the previous lesson, pose this as a whole-class question now.

- Explain that salt and sand is a mixture of two solids, but that the grains in them are of similar size, so it might be difficult to sieve them. Ask the learners: *What happens when sand and salt are mixed with water?*

- Ask the learners to predict with talk partners what happens. Give out photocopiable page 53 for them to record their prediction on.

- Introduce the vocabulary 'soluble' (dissolves in water) and 'insoluble' (doesn't dissolve in water).

Main activities

- Explain that if salt or sand is soluble in water and the other substance isn't, this might give a way to separate them. Ask the learners to think about how this might help separate the substances. Ask them to record their predictions on photocopiable page 53.

- Make sure that there is suitable equipment around the room for the learners to see and think about using. Do not show them how to use any of the equipment available – allow them to choose and try it for themselves. Also allow the

learners to use anything they request that seems reasonable – ask them to justify why they need it and how they are going to use it. Once their method has been approved, they should fill this in on photocopiable page 53. They should then use photocopiable page 54 to record their results and use these to come to a conclusion.

Plenary

- Invite the learners to share what they did. Allow them to demonstrate exactly how they separated the salt solution from the sand.

- Explain that a soluble solid (for example salt) + water makes a solution.

- Remind the learners that substances that dissolve in water are soluble. Make sure that the learners realise that the soluble solid has not disappeared – it is still present in the solution. It is possible to show this by allowing the learners to taste the salt solution (only a drop or two on the tongue and check for allergies first).

- Substances that do not dissolve in water are insoluble.

Success criteria

Ask the learners:

- What is the scientific term for a substance that dissolves in water?
- What do we call a substance that doesn't dissolve in water?
- Is salt soluble or insoluble?
- Is sand soluble or insoluble?
- What do you get when a soluble solid dissolves in water?

Ideas for differentiation

Support: Work in a small group with these learners or organise them to work in mixed-ability groups.

Extension: Ask these learners to think of as many everyday examples of dissolving as they can.

Name: _____

Mixing solids with water 1

Predictions

1. I predict that _____ will be soluble in water.

2. I predict that _____ will be insoluble in water.

Method (what you did)

3. We used _____ of water each time.

4. We used _____ of salt or sand each time.

5. To see if the substance dissolved, we:

_____

_____

_____

6. To separate the insoluble solid from the solution made we
    (write or draw to show what you did):

Name: _____

Mixing solids with water 2

Results (what happened)

1. The _____ was soluble.

2. The _____ was insoluble.

3. We separated them using:

    _____

    _____

Conclusion (what you found out)

4. A mixture of a soluble solid and an insoluble solid can be separated by:

    _____

    _____

    _____

5. Add other substances in the correct columns of this table.

Soluble	Insoluble

Cambridge Primary: Ready to Go Lessons for Science Stage 6 © Hodder & Stoughton Ltd 2013

Soluble or insoluble?

- Explore how, when solids do not dissolve or react with water, they can be separated by filtering, which is similar to sieving. (6Cc4)
- Make a variety of relevant observations and measurements using simple apparatus correctly. (6Eo1)
- Use tables, bar charts and line graphs to present results. (6Eo3)

Resources

Photocopiable page 54 from the previous lesson; filter funnels or cut-off and sandpapered plastic bottles; filter paper or paper towels; sand; water; photocopiable page 56; salt; sugar; flour; coffee; beads; baking powder; chalk; beakers or plastic cups; stirrers; teaspoons; sieves.

Starter

- Ask the learners to share their lists of soluble and insoluble substances on photocopiable page 54 from the previous lesson. Do not comment on their ideas. Explain that they will be able to find out if they were correct in this lesson.
- If you do not have any plastic or glass filter funnels, demonstrate how to make a plastic funnel by cutting the top off a plastic drinks bottle and turning it upside down. Rub the edges with sandpaper to make it smooth to avoid cuts. Alternatively, prepare these before the lesson.
- If you do have filter funnels, demonstrate how to make a filter paper cone. (Fold the circle of filter paper into quarters, then separate one layer from the other three and insert into the funnel.) If you are using plastic bottle-top funnels, use paper towels as filter papers.
- Discuss with talk partners how filter paper works. (It is a very fine sieve.)

Main activities

- Make a mixture of sand and water. Pour it through a filter funnel and paper and open up the filter paper to show the learners the sand remaining on the paper, or use this as an opportunity for the learners to fold filter papers and try for themselves.
- Give out photocopiable page 56 and demonstrate how to complete the table using sand and water as an example. Then ask the learners to choose from the resources available, carry out the activity for each mixture and fill in the table according to what they find. They should then answer question 3.

Plenary

- Discuss the learners' results and correct any misunderstandings or misconceptions.
- Remind the learners that filtration is a scientific process, used to separate a solid from a liquid or to separate a mixture of a soluble and an insoluble solid.

Success criteria

Ask the learners:

- Name a soluble substance.
- Which substances that you tested were insoluble?
- How did you try to make a solution each time?
- What separation method did you use?
- How does a teabag or coffee filter work?

Ideas for differentiation

Support: Assist these learners with folding filter papers and completing photocopiable page 56. Limit their choice of substances to maybe two soluble and two insoluble ones.

Extension: Ask these learners to find out about how a teabag or coffee filter works.

Name: _____

Soluble or insoluble?

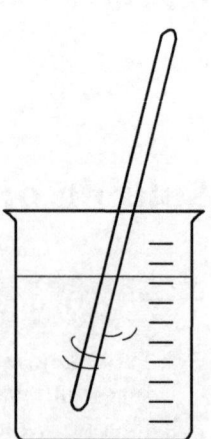

1. Complete the table to show what happened to the substances you tested.

2. In the final column describe anything else that you see, for example colour changes / layers forming.

Substance	Soluble? ✓ or ✗	Insoluble? ✓ or ✗	What else did you see?

3. What happens to a soluble substance when it dissolves in water to make a solution?

_____

_____

   Cambridge Primary: Ready to Go Lessons for Science Stage 6 © Hodder & Stoughton Ltd 2013

Observing solubility

Learning objectives

- Observe, describe, record and begin to explain changes that occur when some solids are added to water. (6Cc3)
- Make a variety of relevant observations and measurements using simple apparatus correctly. (6Eo1)
- Make comparisons. (6Eo4)

Resources

Powder paint; plastic cups or beakers; stirrers; water; chalk; instant coffee; flour; salt; sand; sugar; soil; bicarbonate of soda or any effervescent tablet; interactive whiteboard or flipchart and markers; photocopiable pages 58 and 59.

Starter

- Tell the learners that you are going to mix some powder paint with water. What do they expect to see?
- Make the mixture and ask the learners to tell you what they can see. Alternatively, invite one of them to be a 'magic mixer' and to do this for you! (A coloured solution is formed, which is the same colour as the powder paint that was mixed with the water.)
- Explain that observation is about looking really closely and noticing every detail about what you see. Repeat the solution-making activity with the same or a different colour paint and ask the learners to tell you as soon as they notice something when you are making the mixture. (The paint powder may sit on the surface of the water when you sprinkle it in. The colour may be more intense in one part of the solution until it is thoroughly mixed, and so on.) Make a list of their observations.

Main activities

- Explain that the learners' task in this lesson is to make some more mixtures and to record their observations as carefully as they can. Give out photocopiable page 58 to the learners who need support and photocopiable page 59 to all the other learners. Organise the class into small groups or pairs to carry out the activity.
- Allow them to do the practical work, and travel between groups giving help, support and encouragement by questioning if they are unsure, or challenging them if they find the task easy. Also allow them to try any other mixtures they suggest if it is a reasonable idea.

Plenary

- Explain that solutions can be clear or coloured, but they can also be transparent, translucent or even opaque sometimes.
- Classify the undissolved substances as insoluble.
- Explain that bubbles are evidence of a reaction having taken place. (The bubbles are made of gas.)

Success criteria

Ask the learners:

- Which substances dissolved in water?
- Name something that made a coloured solution.
- Did any of the substances turn solid in water? Which one(s)?
- Which substance(s) did not dissolve?
- Which substance(s) (if any) reacted with the water? How could you tell?

Ideas for differentiation

Support: Give these learners photocopiable page 58 to complete and work from.

Extension: Ask these learners which of the mixtures they made can be separated and how they would do it.

Name: _____

Observing mixtures

You will need:

Plastic cups or beakers, stirrers, water, sugar, sand, flour, powder paint, detergent tablet.

Method (what to do)

● Make mixtures with the substances listed above.

● Record your observations.

Mixture	Soluble? ✓ or ✗	Insoluble? ✓ or ✗	Colour	Fizzing? ✓ or ✗	Layers? ✓ or ✗ What colour?

● Draw an asterix (*) by the name of the mixtures that you could separate.

 Cambridge Primary: Ready to Go Lessons for Science Stage 6 © Hodder & Stoughton Ltd 2013

Name: _____

Observing mixtures

You will need:

Plastic cups or beakers, stirrers, water.

Method (what to do)

- Make mixtures with the substances listed in the table.

- Record your observations.

- Add three more substances of your own choice.

Substance	Soluble? ✓ or ✗	Insoluble? ✓ or ✗	Colour	Fizzing? ✓ or ✗	Layers? ✓ or ✗ What colour?
detergent tablet					
flour					
sand					

- Choose one of the mixtures you could separate and explain how you would do it.

_____

_____

Separating more complicated mixtures

- Explore how, when solids do not dissolve or react with water, they can be separated by filtering, which is similar to sieving. (6Cc4)
- Explore how some solids dissolve in water to form solutions and, although the solid cannot be seen, the substance is still present. (6Cc5)
- Discuss how to turn ideas into a form that can be tested. (6Ep3)

Resources

Sand; sugar; rock salt crystals; beakers or mixing bowls; stirrers; water; sieves; filter papers and funnels; evaporating dishes; photocopiable pages 61, 62, 63 and 64.

Starter

- Show the learners the sand and ask them whether sand is soluble or insoluble.
- Repeat the question for sugar and add the sugar to the sand in a beaker or mixing bowl.
- Do the same for the rock salt crystals and then add these to the mixture of sand and sugar. If the learners are not familiar with rock salt, explain that it is larger crystals of salt.
- Add water to make a mixture of sand, sugar and rock salt.

Main activities

- Explain to the learners that their task in this lesson is to plan, design and carry out a way to separate the mixture back into sand, sugar and rock salt. This is a mixture with more things in it than they have been used to separating previously, so they will have to do it in stages.
- Organise the learners into mixed-ability groups and give out photocopiable pages 61, 62, 63 and 64; one set for each group. Explain the roles and that everyone should take part in some way. Encourage them and give them time to discuss the roles before starting to plan. Set a time limit, for example ten minutes; this will focus them. The awarding of marks is also motivational for some learners.

- This activity may take more than one lesson. This lesson could be used for planning and the next lesson to do the practical work. Look at the photocopiable pages and make sure that the learners know exactly what they have to do before they start.
- Check their plans and approve them, even if they might not be successful.

Plenary

- Discuss either their plans (if the practical is to be done in the next lesson) or their results. If the plans are discussed, do not comment on their proposals – let the learners try the experiment first.
- Invite successful groups to share their method(s).
- Ask less successful groups to comment on where they think they might have done things differently.

Success criteria

Ask the learners:

- What did you do first and why?
- Which substance(s) were insoluble?
- Which substance(s) were soluble?
- What equipment did you use?
- What scientific processes have you used in this experiment?

Ideas for differentiation

Support: Allow these learners to be part of a mixed-ability group. Make sure that they have a clearly defined role within the group and that they participate rather than just watch.

Extension: Ask these learners to think of a different mixture of three things that they could give to another group to separate – but they must know how to do it themselves first.

Name: _____

Separating a complicated mixture 1

1. Decide on a team name for your group – you will be awarded points for this!

    Our team name is: _____

    Points awarded: _____

2. Decide on roles within the group.

    ● **Director** (good at organising):

    [ ]

    ● **Equipment manager(s)** (responsible for making sure that everything is ready when needed):

    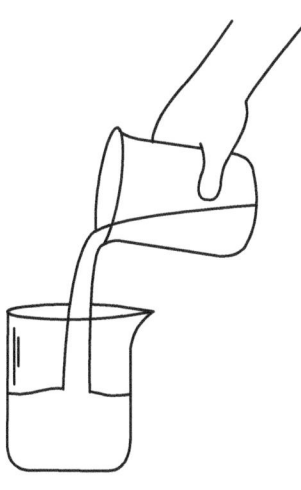

    [ ]

    ● **Recorder** (writes the method, results and conclusion):

    [ ]

    You will be awarded marks out of 10 at the end for how you each performed your role.

Name: _____

Separating a complicated mixture 2

How would you separate a mixture of sand, sugar and rock salt?

Equipment

- _____
- _____
- _____
- _____

Step 1: To separate _____

Draw or write in the box below to show what you will do.

Cambridge Primary: Ready to Go Lessons for Science Stage 6 © Hodder & Stoughton Ltd 2013

Name: _____

Separating a complicated mixture 3

How would you separate a mixture of sand, sugar and rock salt?

Step 2: To separate _____ and _____

Diagram (draw what you would do in the box below)

[ ]

Step 3: To separate _____ and _____

Diagram (draw what you would do in the box below)

[ ]

Name: _____

Separating a complicated mixture 4

> How would you separate a mixture of sand, sugar and rock salt?

Results (what happened)

Write or draw what happened in the box below.

```
[ empty box ]
```

Conclusion (what you found out)

How did you separate a mixture of sand, sugar and rock salt?
Write or draw what you found out in the box below.

```
[ empty box ]
```

Cambridge Primary: Ready to Go Lessons for Science Stage 6 © Hodder & Stoughton Ltd 2013

Investigating solubility

- Observe, describe, record and begin to explain changes that occur when some solids are added to water. (6Cc3)
- Discuss how to turn ideas into a form that can be tested. (6Ep3)
- Make a variety of relevant observations and measurements using simple apparatus correctly. (6Eo1)

Water in beakers at different temperatures (hot and cold tap water, iced water, boiled water from a kettle); sugar or salt; spoons; stirrers; thermometers; room thermometer; stopwatches or timers; photocopiable pages 66 and 67.

Starter

- Ask the learners to discuss with talk partners when solids dissolve in hot liquids on an everyday basis – the learners will probably suggest sugar in hot drinks, detergents in washing machines, bath products and examples from cooking.
- As a class, discuss what happens in each instance to make the solid dissolve – stirring or agitation, usually. Ask the learners to think about what happens to the solid (and powders are very small pieces of solid) when the solution is being made – stirring or agitation breaks up the solid more.
- Prepare a set of liquids or water at different temperatures and ask some of the learners to read the temperature of each using a thermometer. You could use cold water from the tap, iced water from a drinks machine or with ice from the freezer, hot water from the tap, and boiled water from a kettle. Take care to boil the kettle away from the learners and where they will not be in the line of the steam produced when it boils.

Main activities

- Ask the learners: *Do soluble solids dissolve quicker in warmer water? How can we find out?* Listen to the learners' suggestions and lead the discussion to consider how to make the test fair. Consider what would be a suitable volume of water to use and how it can be measured. Also think about the number of stirs and amounts of solid to be used each time. These things are the constant factors. The only factor to be changed should be the temperature of the water. Allow the learners to select which solid to use (either sugar or salt).
- Give out photocopiable pages 66 and 67 for the learners to plan and record their test.
- Allow them to set up and carry out the practical activity and to analyse their results.

Plenary

- Invite different groups of learners to share their results with the rest of the class.
- Consider different sets of results. Do they enable you to reach the same conclusion each time? Is it possible to conclude that the hotter the water, the quicker the solid dissolves?

Ask the learners:

- At which temperature did the solid dissolve quickest?
- How did you make the test fair?
- How many stirs did you do each time?
- Do solids dissolve quicker in hot water?

Support: Organise these learners to work in mixed-ability groups or provide adult support to work specifically with them.

Extension: Ask these learners to also investigate a different liquid (for example milk or fruit juice) and to compare the results with water.

Name: _____

Do solids dissolve quicker in hot water? 1

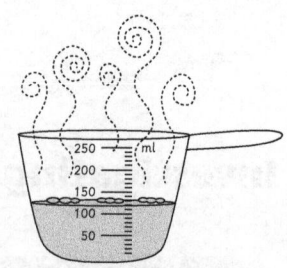

List the equipment you will need:

* _____

* _____

* _____

* _____

* _____

* _____

Method (what you did)

1.   The solid we used was _____.

2.   We used _____ of water each time.

3.   We measured the volume of water using _____.

4.   We used _____ of solid each time.

5.   We measured the amount of solid using _____.

6.   We repeated the experiment using water at different temperatures.

     We measured the temperature of the water using _____.

7.   We timed how long the solid took to dissolve. We used a _____

     to measure the time taken for it to dissolve.

Cambridge Primary: Ready to Go Lessons for Science Stage 6 © Hodder & Stoughton Ltd 2013

Name: _____

Do solids dissolve quicker in hot water? 2

Results (what happened)

Decide the best way to present your results.

Conclusion (what you found out)

The _____ the water, the _____ the solid dissolved.

How can we separate this mixture?

Learning objectives

● Explore how, when solids do not dissolve or react with water, they can be separated by filtering, which is similar to sieving. (6Cc4)

● Explore how some solids dissolve in water to form solutions and, although the solid cannot be seen, the substance is still present. (6Cc5)

● Choose which equipment to use. (6Ep7)

Resources

Internet access; photocopiable pages 69 and 70; muddy water from a pond; saturated salt solution; evaporating dishes or saucers; heat source; matches; soil; sieves; filter funnels and filter papers; glass beakers or jars; stirrers; plastic teaspoons; water; reference materials.

Starter

• Go to www.scibermonkey.org → ages 7–11 → Materials and changes → Separating solids and liquids. Select 'more sites' and look at the 'Separating materials' film clip. Also look at the 'Changes can be temporary or permanent – part 2' activity on the same page of the site. Both provide good reminders and revision about separating mixtures.

• Ensure that you are confident that the learners have sufficient knowledge to undertake one or all of the activities in this lesson.

Main activities

• Explain that today the learners will try to separate different types of mixtures by doing several different activities (or one – depending on the class organisation you have chosen). There are three activities – cleaning dirty water, growing crystals and separating soil.

• Give out photocopiable pages 69 and 70 and explain what each task is about.

• Organise the class into pairs or small groups and allocate them an order in which to carry out each activity. Allow the learners to do the practical work, then bring them back together as a whole class.

• Alternatively, depending on the availability of resources, split the class into three groups and plan for each group to do one of the activities, then feed back to the rest of the class during the Plenary session.

Plenary

• Go back to the same Scibermonkey page used during the Starter activity, click on 'Types of change' and do the quiz. This could be introduced with a competitive element, for example one half of the class against the other, or between table groups. Award points and maybe a small prize for the winning team.

Success criteria

Ask the learners:

● How did you separate the dirty water?

● Will crystals be bigger or smaller after slow evaporation?

● How many different sizes of particles did you find in soil?

● Which separation processes are involved in separating dirty water?

● Which scientific process is similar to sieving?

Ideas for differentiation

Support: Organise these learners to work in mixed-ability groups or provide adult support to work with them as a small group. Alternatively, allocate the separating soil activity to this group if only one activity is to be carried out in groups.

Extension: Ask these learners to find out about the production of salt on an industrial scale and what processes of separation are involved.

Name: _____

Separating different mixtures 1

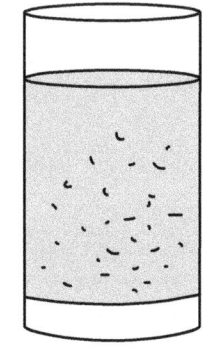

Show your teacher your ideas before you begin.

---

1. **Making dirty water clean**

You will need:

Some dirty water.

- What is making the water dirty?

  _____

- What equipment will you need?

  _____

  _____

- How exactly will you make the water clean?
  (Clue: it might take one or more steps.)

  _____

  _____

---

2. **Making salt crystals**

You will need:

Saturated salt solution.

- How can you get the salt out as crystals?

  _____

- What equipment will you need?

  _____

  _____

- How exactly will you do it?

  _____

  _____

Name: _____

Separating different mixtures 2

3. **Separating soil**

You will need:

Some soil.

● How will you separate the soil?

_____

_____

● What equipment will you need?

_____

_____

● How exactly will you do it?
(Clue: it might take one or more steps.)

_____

_____

_____

What processes have you used today?

Tick (✓) those you have used from this list:

● condensing ☐

● evaporating ☐

● filtering ☐

● sieving ☐

Unit assessment

- What kinds of change are reversible? (Physical changes.)
- Are chemical changes reversible or irreversible? (Irreversible.)
- How would you separate a mixture of rice and dried peas? (Sieving.)

- What do soluble solids do when they are mixed with water? (Dissolve.)
- What do we call solids that do not dissolve in water?
- Name a process that is similar to sieving. (Filtering.)

Summative assessment activities

Observe the learners while they participate in these activities. You will quickly be able to identify those who appear to be confident and those who may need additional support.

Choosing the right equipment

This activity assesses the learners' ability to choose the right equipment.

You will need:

A selection of equipment used to separate mixtures; sieves, filter papers and funnels, evaporating dishes.

What to do

- Give the learners an example of a mixture and ask them to show and tell you how they would separate that mixture, for example lentils and sand, flour and sugar, sugar and water, coffee and water. Use examples of everyday substances, or substances that the learners have worked with in class.
- Record their responses on a class spreadsheet or checklist.

Separating mixtures

This activity assesses the learners' ability to separate mixtures.

You will need:

A mixture of chalk, iron filings and salt; magnets; water; beakers or plastic cups; stirrers; heat source.

What to do

- Tell the learners what is in the mixture and ask them in pairs to choose from the equipment provided to separate the three components.
- Observe and record their method and ask questions about what processes are involved as they work through the task.

Distribute photocopiable pages 72, 73 and 74. The learners should work independently, or with the usual adult support they receive in class.

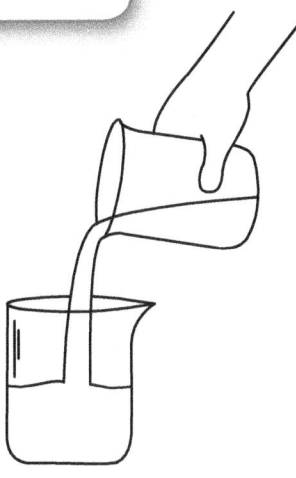

Name: _____

Interpreting results: melting ice cubes

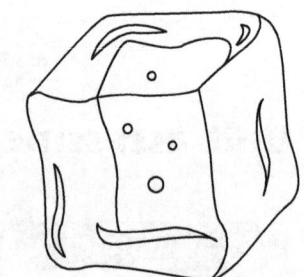

Class 5 have been melting ice cubes in different places.

Here are their results.

Look at the table and answer the questions below.

	Room temperature (°C)	Time taken to melt completely (minutes)
Place A	20	21
Place B	18	23
Place C	17	25
Place D	10	30

1. In which place did the ice cube melt quickest? _____

2. Explain why the ice cube melted quickest here.

    _____

    _____

3. Which was the coldest temperature recorded? _____

4. Use these results to draw a bar chart in the box below.

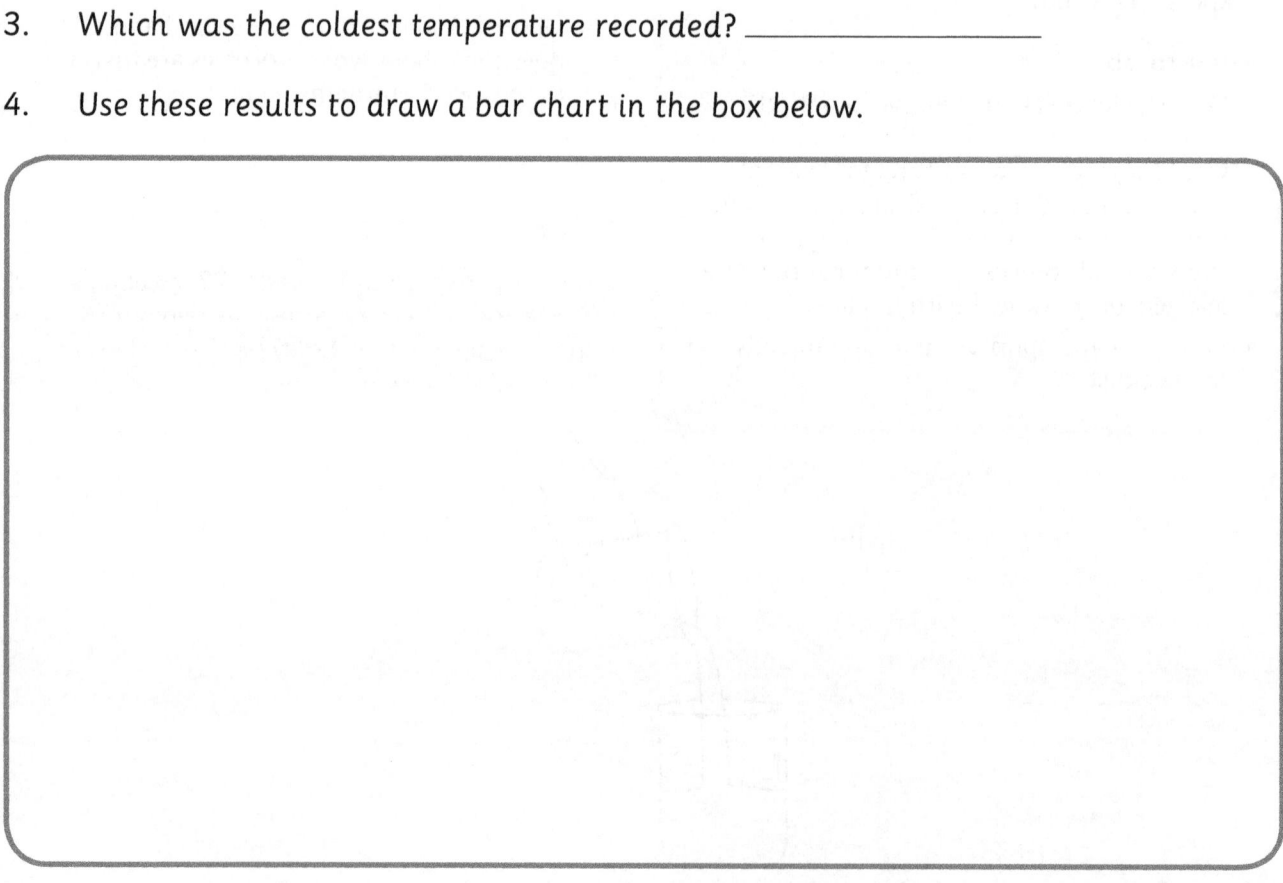

Cambridge Primary: Ready to Go Lessons for Science Stage 6 © Hodder & Stoughton Ltd 2013

Name: _____

Separating mixtures

1. Give the names of the process or processes you would
    use to separate the mixtures listed below.

    Use these words to help you.

    | by hand | evaporation | filtration | magnetism | sieving |

    a) Iron filings and chalk: _____

    b) Glass marbles and beads: _____

    c) Soil and gravel: _____

    d) Sand and water: _____

    e) Salt and water: _____

    f) Rice, water and sugar: _____

2. What equipment would you use to separate what is in soil?

    _____

3. What equipment would you use to separate iron filings and sand?

    _____

4. What equipment would you use to separate salt from salt water?

    _____

Reversible or irreversible?

Tick (✓) to show whether these changes are reversible or irreversible.

Change	Reversible?	Irreversible?
water boiling		
a candle burning		
chocolate melting		
an egg cooking		
water freezing		
a fire burning		

 Cambridge Primary: Ready to Go Lessons for Science Stage 6 © Hodder & Stoughton Ltd 2013

Habitats and living relationships

Learning objectives

● Know how food chains can be used to represent feeding relationships in a habitat and present these in text and diagrams. (6Be3)

● Make comparisons. (6Eo4)

Resources

Internet access or reference materials; interactive whiteboard or flipchart and markers; photocopiable page 76.

Starter

• Ask the learners to discuss with talk partners: *What is a habitat?* Listen to their responses and agree to a definition of a habitat as the environment in which an animal or plant can normally be found – its natural environment.

• Show some pictures of different habitats from the internet or reference books and ask the learners to identify them, for example a desert, a mountain, the ocean, and so on.

• Introduce two new terms. First, all the animals and plants that exist and live together in a particular habitat are a **community** – like a local community of people in a neighbourhood, town or village. Ask the learners to name some animals and plants that would commonly be found in the desert (for example camels, desert rats, cacti, and so on). Compare different animals and plants found in different habitats.

• Second, explain that within a habitat sometimes there are animals or plants of the same species, for example different types of ferns or birds. These are known as a **population**. Ask the learners to identify different species of monkeys that might be found in a rainforest (gorillas, chimpanzees, orang-utans, gibbons, and so on). Compare different species within particular populations.

Main activities

• Explain that all living things need energy for growth. Ask the learners: *Where do animals and plants get their energy from?* (Animals from their food and plants from the Sun.)

• When an animal moves, it uses up energy. Ask the learners: *What actions or movements have you done so far today?* Compile a class list of responses and discuss the many things that their bodies have needed energy for.

• Give out photocopiable page 76 and ask the learners to complete it. Explain that they need to think about where they get their energy from, and try to work out where the ultimate source of their energy comes from.

Plenary

• Explain that the process of obtaining energy is known as a **food chain** and can be represented in a particular way, which the learners will find out about in the next few lessons.

Success criteria

Ask the learners:

● What do all animals and plants need to grow, move and survive? (Energy.)

● Where do plants get their energy from? (The Sun.)

● What is the source of energy for humans? (Food.)

● What is the preferred, natural environment for a particular plant or animal known as? (A habitat.)

● What is a population?

● What is a community?

Ideas for differentiation

Support: Work alongside these learners as they complete photocopiable page 76.

Extension: Ask these learners to choose an animal or plant and trace back its energy to the ultimate source of energy. Use the same format as photocopiable page 76.

Name: _____

Energy source

1. Draw a picture of yourself.

[box]

2. Name **one** food that gives you energy. _____

3. Where does this food get its energy from? Draw or write your answer in the box below.

[box]

4. What is the ultimate source of energy? _____

Cambridge Primary: Ready to Go Lessons for Science Stage 6 © Hodder & Stoughton Ltd 2013

Food chains

Learning objectives

- Know how food chains can be used to represent feeding relationships in a habitat and present these in text and diagrams. (6Be3)
- Make comparisons. (6Eo4)

Resources

Photocopiable pages 76, 78 and 79.

Starter

- Re-cap on learning from the previous lesson and ask the learners to discuss with talk partners: *What is the ultimate source of energy for animals and plants?* (The Sun.) Use photocopiable page 76 for reference.
- Demonstrate how to complete photocopiable page 76 by using an example given by a learner who needs support. Explain how to start with the producer and use arrows. In general:

  green plant → animal that produces meat → human.

- Give out photocopiable page 78 and allow the learners to use their information from photocopiable page 76 to help them complete photocopiable page 78. Explain that the top example is for meat / fish-eaters (three boxes) and the second example (two boxes) is for vegetarians or non-meat-eaters.

Main activities

- Ask the learners to refer to their completed photocopiable page 78 throughout the rest of the lesson. Work through the general outline of a food chain, as presented in the Starter activity. Ask the learners for specific examples from their completed photocopiable page 78.
- Make sure that the learners all understand that the Sun is the ultimate source for all food chains. The first stage in a food chain is always a green plant, then an animal, then humans (or simply a green plant then humans for vegetarians). Compare the difference between a food chain drawn for a vegetarian and a non-vegetarian.

- Ask the learners to think about what the arrows mean. Explain this as 'gets energy from' or simply 'eats'. Use an example:

  leaf → caterpillar → bird → cat

  and explain this as the caterpillar eats the leaf, the bird eats the caterpillar and the cat eats the bird.

  Or

  The caterpillar gets its energy from eating the leaf, the bird gets its energy from eating the caterpillar and the cat gets its energy from eating the bird.

- Give out photocopiable page 79 and ask the learners to complete the food chains on this page. Explain that it is very important to draw the arrows in the correct direction, as shown above.

Plenary

- Explain that food chains show feeding relationships in a particular habitat. Ask the learners to describe, using the arrows, the flow of energy in any food chain.
- Make sure that they realise the importance of green plants in food chains.

Success criteria

Ask the learners:

- What does a food chain represent?
- What do the arrows represent? (The flow of energy.)
- What do all food chains begin with? (A green plant.)
- What is the next thing in a food chain?

Ideas for differentiation

Support: Assist these learners in completing photocopiable page 79.

Extension: Ask these learners to draw a food chain for a particular habitat that is familiar to them locally.

Name: _____

Food chains

1. Use the information from the 'Energy source' page to help you draw a food chain below. Draw in the boxes and write on the lines provided.

Meat-eater

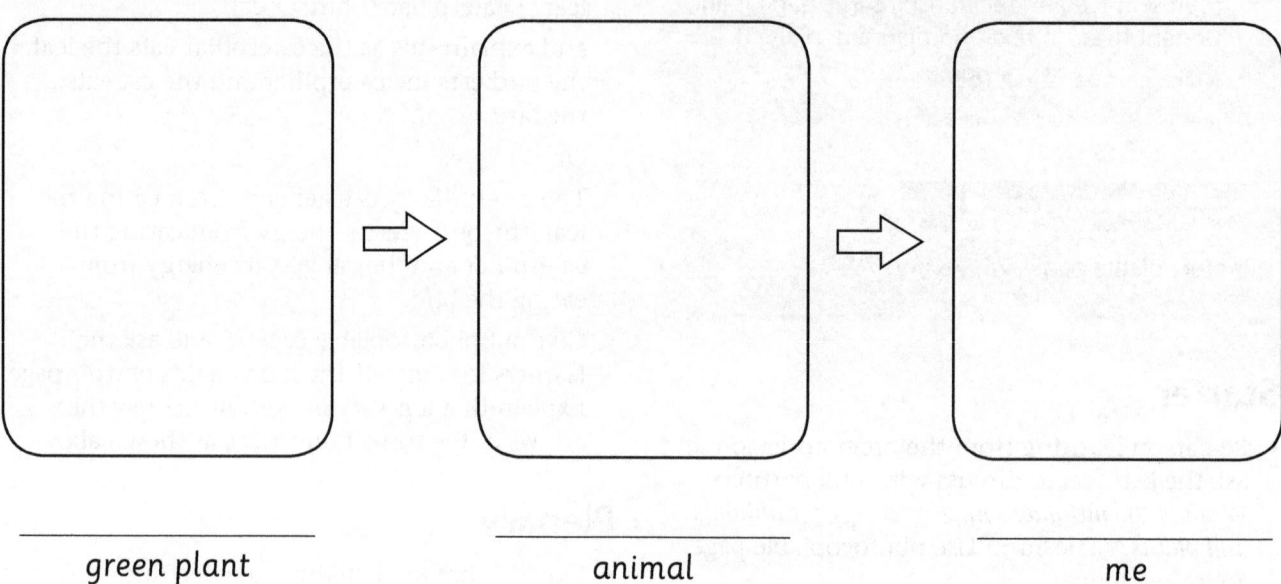

green plant	animal	me

Vegetarian

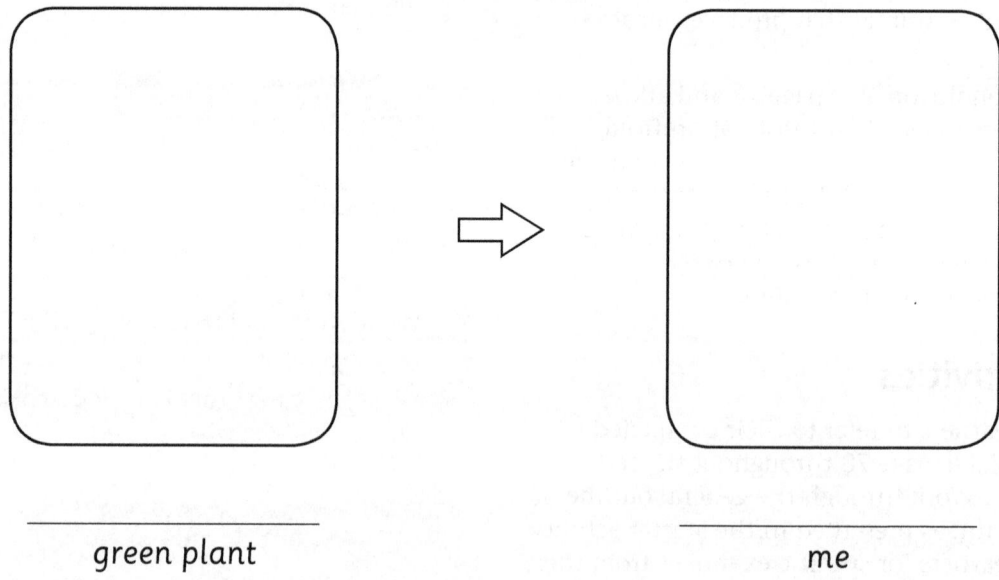

green plant	me

2. Compare the differences and similarities between these two food chains.

   What do you notice?

   _____

   _____

 Cambridge Primary: Ready to Go Lessons for Science Stage 6 © Hodder & Stoughton Ltd 2013

Name: _____

Drawing and writing food chains

1. Draw arrows to complete this food chain.

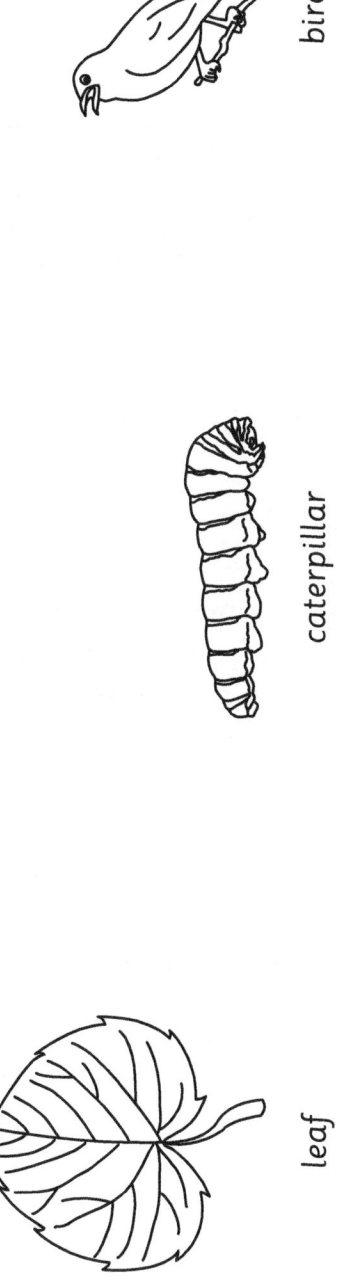

leaf

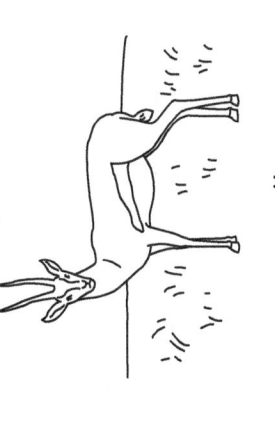

caterpillar

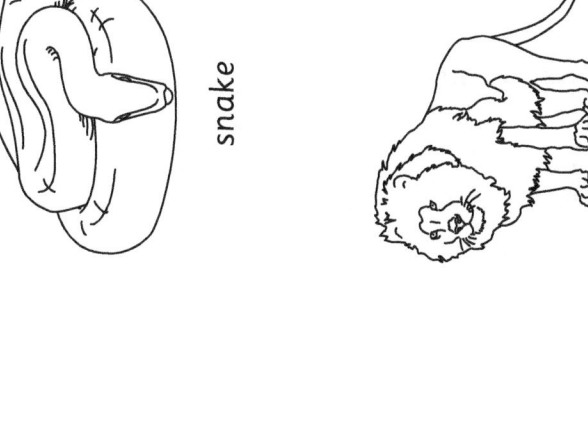

bird

snake

2. Complete the blanks in this food chain. (Remember to add the arrows!)

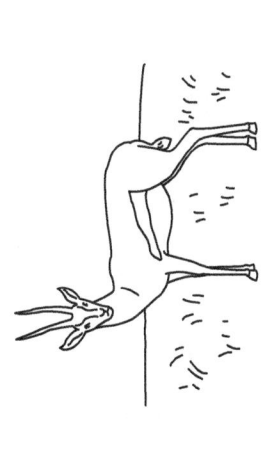

gazelle

_____

_____

3. What do the arrows in a food chain show?

_____

_____

Food chains and producers

● Know that food chains begin with a plant (the producer), which uses energy from the sun. (6Be4)

● Make comparisons. (6Eo4)

Green plants; sticky notes; photocopiable page 81; internet access or reference materials; paper, art materials for poster-making or an ICT package.

Starter

• Ask the learners to name the process by which green plants are able to make their own food (photosynthesis).

• Explain that because green plants are the only living things to do this (humans and animals can't do this – we have to consume our food), plants are called **producers**.

• Give the learners in pairs or small groups a green plant and some sticky notes. Ask them to use the green plant and sticky notes to write about what happens during photosynthesis. (This is good revision from Stage 5.) Ask them to use the sticky notes as labels on and around the plant. Allow them to write or draw their responses. Provide an opportunity for each group to show the rest of the class how they have done this. Put right any misunderstandings or misconceptions.

Main activities

• Give out photocopiable page 81 and explain that there are several food chains on this page. Ask the learners to consider which particular green plant could be the producer for each food chain.

• Show the learners the reference materials available, or allocate time slots for using the internet if it is available. Alternatively use an ICT suite if facilities allow.

• Organise the learners into pairs or small groups for this activity, depending on the availability of resources. Choose whether you want them to work in ability or mixed-ability groups. This will affect how you distribute resources. Alternatively, give the learners who need support fewer examples to complete within the same lesson.

Plenary

• Make sure that the learners remember what happens during photosynthesis. Write the following word equation on the board:

carbon dioxide + chlorophyll + water + light → starch (food) + oxygen

Remind the learners or ask them to describe what happens by explaining what this word equation means. Alternatively, ask a learner to write the equation and explain it.

• Explain that there are some insect-eating plants, for example Venus fly-trap – but these are an exception.

• Go through the responses to the food chains on photocopiable page 81.

• Compare producers in each food chain.

Ask the learners:

● What do all food chains begin with?

● Why are green plants known as producers?

● What is the process by which green plants make their food?

● Where do green plants get their energy from?

Support: Organise these learners to work in a mixed-ability group for the Starter activity.

Extension: Ask these learners to make a poster showing what happens during photosynthesis.

Name: _____

Producers

1. Name or draw a suitable producer in each of these food chains.

a)

[box]  bird  owl

_____
producer

b)

[box]  sheep  wolf

_____
producer

c)

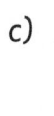 [box]  chicken  snake

_____
producer

2. How could you group these producers?

They are all _____.

Consumers

Learning objectives

● Understand the terms *producer, consumer, predator* and *prey.* (6Be5)
● Make comparisons. (6Eo4)

Resources

Cards prepared from photocopiable pages 83 and 84 (or labels with the words 'carnivore', 'herbivore' and 'omnivore' written on them); internet access or reference materials – pictures of animals that are herbivores / carnivores / omnivores; flipchart and markers or whiteboard; photocopiable page 85.

Starter

- Organise the learners to play a game in pairs or small groups: Before the lesson, prepare sets of cards using photocopiable pages 83 and 84. Ask the learners to group the pictures of the animals on the cards into herbivores (eat plants), carnivores (eat meat) and omnivores (eat plants and meat). Alternatively, prepare some cards with the terms on them and use some pictures from the internet or books, for example pictures of animals such as cattle, horse, grasshopper (herbivores); lion, eagle, shark, spider (carnivores); human, bear, raccoon (omnivores).

- You may need to discuss these words with the learners before they start. If necessary, display definitions for them to refer to while doing the activity.

- For differentiation, you could give groups the same pictures, but ask them to categorise different numbers of pictures. Alternatively, select different pictures for each group – simple, familiar animals for the learners who need support and rarer animals for the learners who usually need to do extension activities.

Main activities

- Explain that the next stage in a food chain after the producer is a consumer. To consume means 'to eat'. Therefore, herbivores are the next link in a food chain, and they are known as 'primary consumers' because they are the first consumer in the food chain.

- Give out photocopiable page 85 and explain that the learners have to identify the primary consumers in the food chains on this page.

Plenary

- Play the Starter activity game again as a class. Give the learners other examples to include, or ask those who have done the extension activity to show what they have also included. Try to make the lists as long as you can.

- Make sure that the learners know the difference between the terms **herbivore**, **carnivore** and **omnivore**. Compare and discuss the differences between each of these groups of animals and why they eat what they eat.

- Ask the learners to make subgroups of herbivores, carnivores and omnivores if possible. This could include, for example, a group of birds of prey such as eagle, hawk and owl in the carnivore group, and so on.

Success criteria

Ask the learners:

● What is a herbivore?
● What do carnivores generally eat?
● Name another omnivore besides humans.
● What does 'to consume' mean?
● Why are herbivores primary consumers?

Ideas for differentiation

Support: Give these learners fewer cards to group when playing the game in the Starter activity.

Extension: Ask these learners to draw and add more herbivores, carnivores and omnivores to the groups in the Starter activity.

Consumers

herbivores

carnivores

omnivores

human

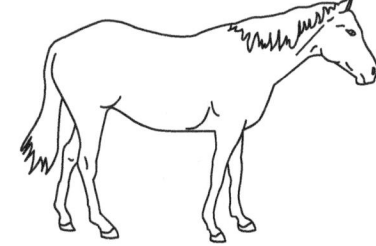

horse

polar bear

tiger

Consumers

eagle

shark

chimpanzee

crocodile

frog

lizard

giraffe

buffalo

dolphin

goat

 Cambridge Primary: Ready to Go Lessons for Science Stage 6 © Hodder & Stoughton Ltd 2013

Name: _____

Consumers

1. Underline the primary consumer in these food chains.

a)

grass   bird   owl

b)

leaves   goat   girl

c)

leaves   antelope   eagle

2. Draw a food chain of your own and label all the animals and plants in it.

   Circle the primary consumer.

_____ _____ _____

Secondary consumers

Starter

- Ask the learners to bring or give out again their finished copies of photocopiable pages 78, 79, 81 and 85. Ask a range of questions about these food chains as revision of what has been learnt so far in this unit, for example: *Name a producer from any of the food chains. What do all producers have in common? Name an animal that eats grass. What does a bird eat?*

- Encourage the learners to give answers other than those shown in the food chains on the photocopiable pages being used for reference for this activity.

- Ask the learners to identify with talk partners the primary consumer in each of the food chains on the completed photocopiable pages. Listen to and discuss their responses and make a class list of them. Display this for reference during the rest of the lesson.

- Revise how a producer is always a green plant and the thing that eats the producer is always the primary consumer.

Main activities

- Explain that in this lesson the learners will find out about the next steps of a food chain.

- Give the learners a short time limit (for example 20 seconds) to discuss with talk partners what the next thing after the primary consumer in a food chain might be called.

- Listen to their responses and introduce the new vocabulary: **secondary consumer**.

- It might be necessary to explain the terms **primary**, **secondary** and **tertiary** if these are unfamiliar terms to the learners. Simply explain it as primary = first, secondary = second, and tertiary = third. (Tertiary might only need explaining to the learners who do the extension activity.)

- Select any of the food chain examples already used on previous photocopiable pages and ask the learners to identify the secondary consumers on each of them.

- Give out photocopiable page 87 and explain that the learners have to complete, label, draw or write food chains.

- Give out photocopiable page 88 to the learners who need extension after they have completed photocopiable page 87.

Plenary

- Invite some of the learners to share the food chains that they have constructed. Consider if they are correct; if not, ask questions to help them to change their answer.

- Use examples to ask the learners which are secondary consumers.

Name: _____

Secondary consumers

A **secondary consumer** consumes (eats) the primary consumer in a food chain.

> **Remember!**
>
> The **producer** is always a green plant.
>
> The **primary consumer** is the thing that consumes (eats) the producer.

1. Complete this food chain by adding a **secondary consumer** of your choice.

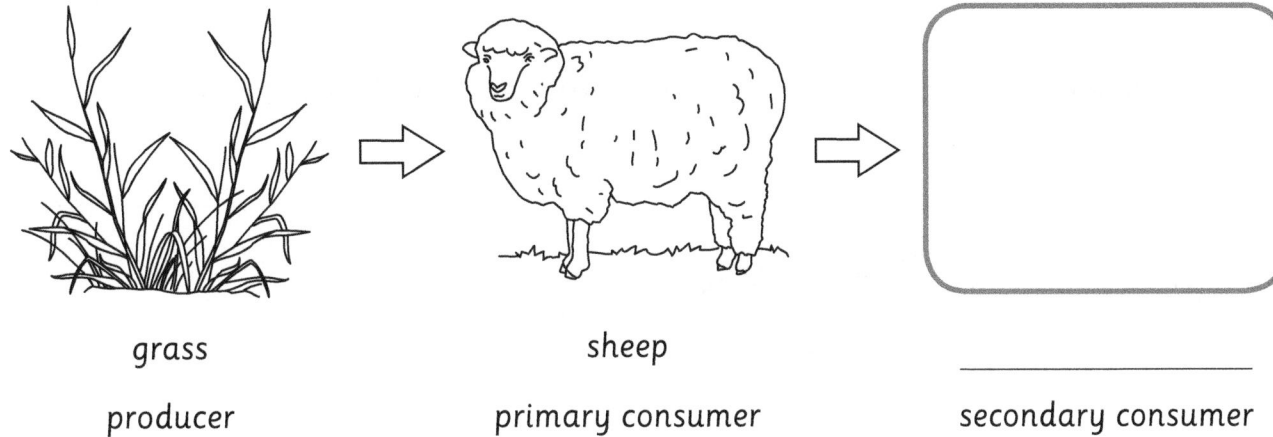

grass	sheep	_____
producer	primary consumer	secondary consumer

2. Label this food chain using the words below (use some words in pairs).

> consumer   primary   producer   secondary

          leaves   caterpillar   bird

_____   _____   _____

_____   _____

3. On the back of this page, draw and write a labelled food chain of your own.

Tertiary consumers

A **tertiary consumer** consumes (eats) the secondary consumer in a food chain.

> **Remember!**
>
> The **producer** is always a green plant.
>
> The **primary consumer** is the thing that consumes (eats) the producer.
>
> A **secondary consumer** consumes (eats) the primary consumer in a food chain.

1. Complete this food chain by adding a **tertiary consumer** of your choice.

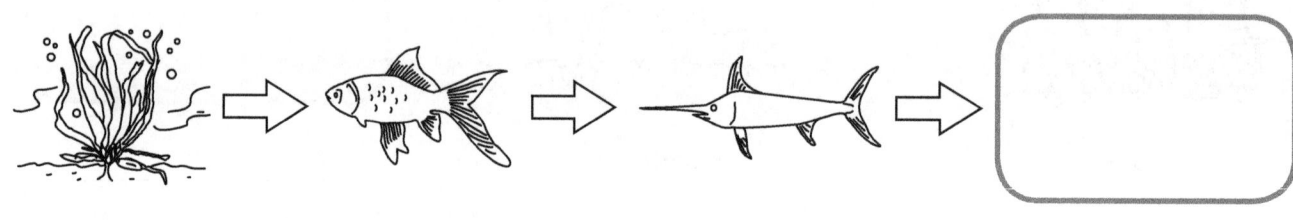

| seaweed | small fish | big fish | _____ |
| producer | primary consumer | secondary consumer | tertiary consumer |

2. Complete these sentences about food chains. Use the following words to help you. (You might need to use some of them more than once, or not at all.)

> carnivores   green plant   herbivores   omnivores   Sun

a) The ultimate energy source for all food chains is the _____.

b) Food chains always begin with a _____.

c) Primary consumers are usually _____.

d) Secondary consumers can be _____ or

_____.

e) Tertiary consumers are always _____.

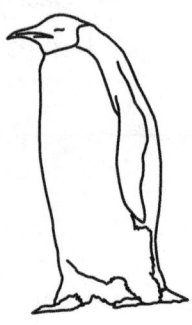

 Cambridge Primary: Ready to Go Lessons for Science Stage 6 © Hodder & Stoughton Ltd 2013

Predators

- Know how food chains can be used to represent feeding relationships in a habitat and present these in text and diagrams. (6Be3)
- Understand the terms *producer, consumer, predator* and *prey*. (6Be5)
- Make comparisons. (6Eo4)

Resources

Previous examples of food chains; photocopiable page 90; internet access or reference materials; sticky labels with the word 'predator' written on them; pictures of animals and plants.

Starter

- Ask the learners to think with talk partners of words that could describe any consumer in a food chain (primary, secondary or tertiary). Remind the learners that each of these things consume (eat) something else lower down the food chain.
- Introduce the word **predator** and describe it as something that eats or predates on something else.
- Look back over food chains that have been drawn, written or looked at in this unit. For each food chain, identify what each consumer is the predator of. Alternatively, use examples from the internet or reference books. Use the sticky labels to label the predators. Continue with these examples until you are sure that the learners understand the term **predator**.

Main activities

- Give out photocopiable page 90 and explain that the learners have to identify the predators in each of the food chains shown. Explain that there will be more than one predator in each food chain, because there are several consumers, which all consume something else.

- Go over their responses, correcting where necessary and addressing any misconceptions or misunderstandings.
- Identify whether each of the named predators is a herbivore, carnivore or omnivore.

Plenary

- Play a game: Give the learners pictures of animals and plants for them to design their own food chains. Invite them in groups to make a food chain, then present it to the rest of the class, explaining which things in it are predators and what they predate on.
- Discuss what would happen if something in the environment affected the numbers of a certain predator (for example if millipedes died out, there would be fewer birds, as they would not have enough to eat, but there would be more small insects as the millipedes would not be there to eat them).

Success criteria

Ask the learners:

- What is a predator?
- Is there only one predator in a food chain?
- Give a reason for this.
- What would happen to the rest of the animals in a food chain if one of the predators died off?
- Tell the rest of the class something that a particular predator predates on.

Ideas for differentiation

Support: Assist these learners during the Plenary game by working with them in a small group, or organising them into mixed-ability groups.

Extension: Ask these learners to choose one predator and research exactly what it likes to eat. Alternatively, give each of them the name of a specific predator to research.

Predators

Predators are things in a food chain that consume (eat) something else.

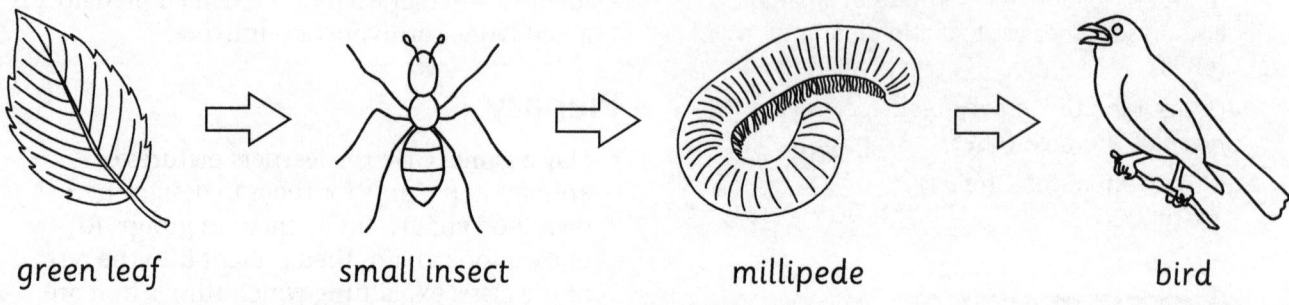

| green leaf | small insect | millipede | bird |

1. List the predators in the food chain above.

    ● _____

    ● _____

    ● _____

2. Which predator do you think there will be most of in this food chain?

    _____

3. Which predator do you think there will be least of in this food chain?

    _____

4. Explain your answer to question 3.

    _____

    _____

    _____

5. Name some predators you might find in the jungle.

    _____

    _____

    _____

 Cambridge Primary: Ready to Go Lessons for Science Stage 6 © Hodder & Stoughton Ltd 2013

Prey

- Know how food chains can be used to represent feeding relationships in a habitat and present these in text and diagrams. (6Be3)
- Understand the terms *producer, consumer, predator* and *prey*. (6Be5)
- Make comparisons. (6Eo4)

Cards with a horizontal arrow on; pictures of a range of plants and animals from a variety of habitats; digital camera(s); photocopiable page 92; internet access and interactive whiteboard or screen.

Starter

- Prepare some cards with a horizontal arrow on. Also provide a selection of pictures of plants and animals from different habitats. Organise the learners into small groups and ask them to make as many different food chains as they can with the pictures and arrow cards that they have been given. Set a time limit, for example two minutes.
- Invite them to share their food chains with the rest of the class, or to choose a particular one to share.
- Use this activity as an opportunity to check the learners' understanding of relevant vocabulary by asking questions about producers, consumers and predators as they show their finished work.
- Allow them to take digital photographs as a record of this work if digital cameras are available. The photographs can then be inserted into their Science workbooks as evidence of the activity.

Main activities

- Select one food chain from each small group and ask the learners to identify the predators.
- Introduce the new term **prey** to classify things that are predated on by predators.

- Ask the learners to identify the prey in their food chains. Make up some different food chains using the pictures available.
- Give out photocopiable page 92 for the learners to record their food chains on and to think of more examples.

Plenary

- Watch some film clips of predators and prey at www.youtube.com: Search for the clip by the title 'predators and prey' or search by film clip reference number GFTBFDPfMh8. This is a good clip as it shows the chase, but not the killing. However, there are many good clips available here – view and choose the most appropriate one(s) for your learners.
- Discuss the clips shown, identifying the prey in each instance. The discussion might lead on to how there can be more than one type of prey for a predator.

Ask the learners:

- What is prey?
- What is a predator?
- Are there more predators than prey in a food chain?
- Name a predator of sheep.
- What could be prey for a cheetah?

Support: Be selective and provide these learners with pictures of familiar animals and plants to construct their food chains from.

Extension: Give these learners (or ask them to choose) an example of a habitat. Ask them to prepare a list of predators and prey that might be found there, for example a desert, the ocean.

Name: _____

Prey

1. Use the pictures and arrows from the Starter activity to **write** a food chain that you made. Label the prey.

   Here is an example:

   grass $\Rightarrow$ rabbit $\Rightarrow$ fox

   prey

My food chain

2. Now make up two more food chains and label them in the same way.

a)

b)

3. Are there more predators than prey in a food chain? yes / no

Cambridge Primary: Ready to Go Lessons for Science Stage 6 © Hodder & Stoughton Ltd 2013

Predators and prey

Learning objectives

● Know how food chains can be used to represent feeding relationships in a habitat and present these in text and diagrams. (6Be3)

● Understand the terms *producer, consumer, predator* and *prey*. (6Be5)

● Make comparisons. (6Eo4)

Resources

Internet access and whiteboard or screen; photocopiable pages 94 and 95; scissors; felt-tipped pens or colouring pencils; card; glue; a laminator and laminating pouches.

Starter

• Show some film clips from the internet of predators and prey, as in the Plenary activity from the previous lesson. Make sure that you select the clips carefully so that they will not cause distress among the learners – some of them will be quite sensitive regarding animals being hurt.

• Discuss the clips either during or straight after viewing them and concentrate on which animals are the predators and which are the prey each time. Introduce the concept of prey being predated on by predators and make sure that the learners can confidently use the terms **predator** and **prey**.

Main activities

• Prepare the cards from photocopiable pages 94 and 95: Copy them onto card, colour them in if preferred and laminate them for future use if a laminator is available.

• Play the game as described on photocopiable page 94. The learners have to classify the animals as predators or prey; if the animal is a predator, the learners have to say what its prey is. If the learners identify the animal as prey, they have to say what its predator(s) might be. A correct response means that the learner keeps the picture. The winner of the game is the learner who has the most cards when all the cards have been collected.

• Award a small prize – perhaps one of the pictures of the animals – to the winning learner.

Plenary

• Use the pictures again and ask the learners to name something that the animal preys upon or is its predator.

• Explain that in a particular habitat, many different food chains exist, because of the variety of plants and animals in it. Some predators have different types of prey depending on the time of year (for example in a pond, larvae are only present for a short time each year before they grow into adults).

Success criteria

Ask the learners:

● Are predators producers or consumers?

● Is prey a consumer or producer?

● What is the last thing in any food chain – a producer, consumer, predator or prey?

● A lion is a predator. What might it choose as its prey?

● A water snail can be prey. What eats water snails in a pond?

Ideas for differentiation

Support: Give these learners fewer cards to use in their game and have a separate winner from this group also.

Extension: Ask these learners to devise a table to show the predators and prey that they identify during the game.

Predators and prey 1

Play this game to identify animals as predators or prey.

How to play:

1. Mix up the cards.

2. Place them in a pile, face down in the centre of the table.

3. Take turns.

4. Turn over a card and say if the animal can be a predator or prey.

5. If it is a predator, tell the other players what that predator eats.

6. If it is prey, tell the other players which animal or animals might prey on it.

7. If you give a correct response, keep the card.

8. The next player then takes a turn.

9. Continue until all the cards are used up.

10. The winner is the player with the most cards.

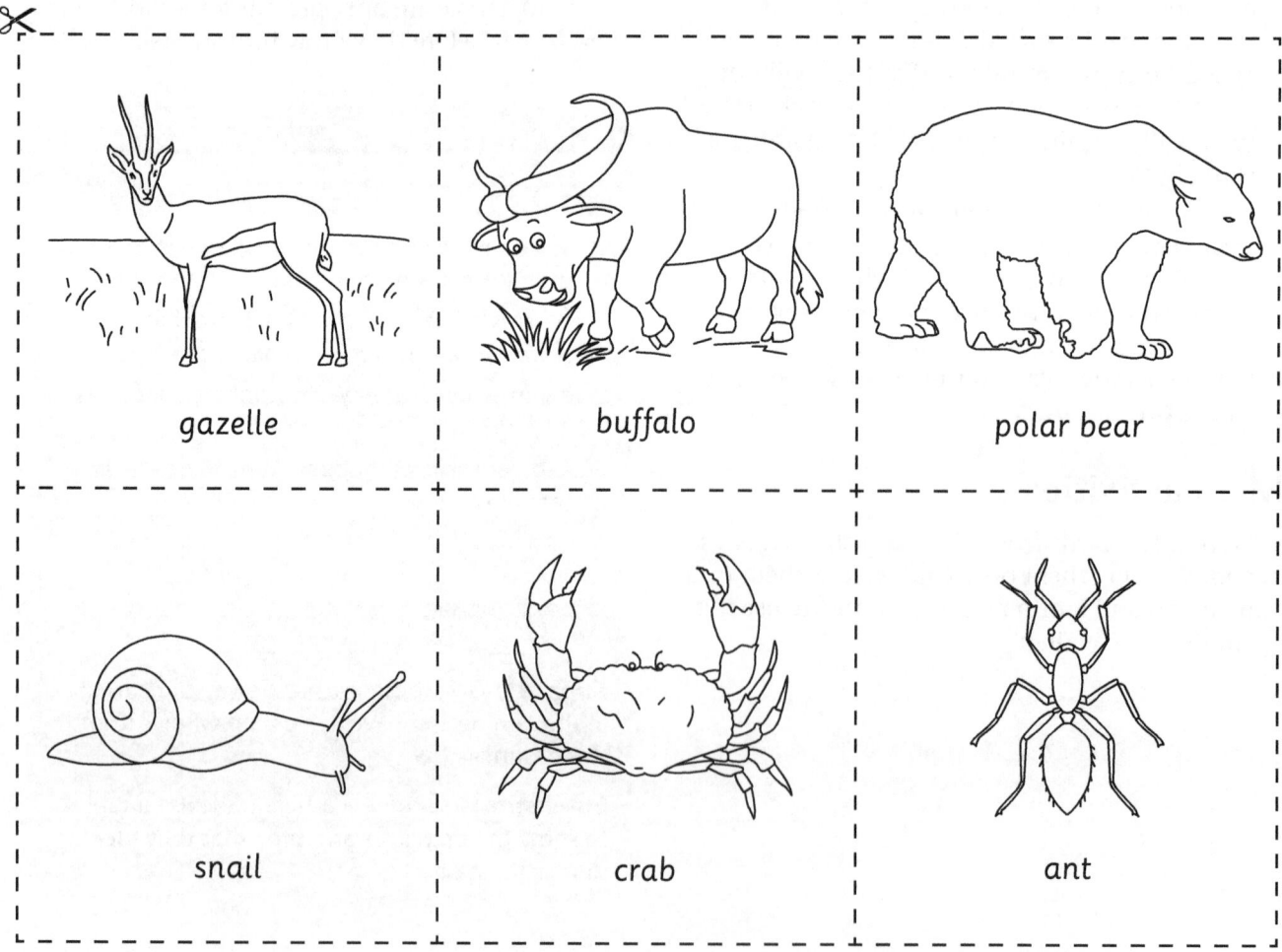

gazelle	buffalo	polar bear
snail	crab	ant

Cambridge Primary: Ready to Go Lessons for Science Stage 6 © Hodder & Stoughton Ltd 2013

Predators and prey 2

snake

shark

crocodile

tiger

fish

goat

wolf

eagle

caterpillar

sheep

chicken

frog

More food chains

- Understand the terms *producer, consumer, predator* and *prey.* (6Be5)
- Explore and construct food chains in a particular habitat. (6Be6)
- Make comparisons. (6Eo4)

Card game from the previous lesson; pictures of animals and plants used previously in this unit; arrow cards; photocopiable pages 97–104; scissors; glue; flipchart and markers or whiteboard; globe or world map.

Starter

- As a class or in small groups of learners, play the game again from photocopiable pages 94 and 95.
- Introduce pictures of plants as producers, and ask the learners to identify from the pictures available what the primary consumer(s) of that producer might be.
- Give the learners in small groups a selection of arrow cards and pictures of animals and plants and ask them to construct a food chain.
- Ask each group in turn to show the rest of the class the food chain that they have constructed. Invite the rest of the class to question the group about their food chain.

Main activities

- Explain that in this lesson and the following few lessons the learners will look at food chains in different habitats around the world, using a different photocopiable page each time from photocopiable pages 97–104.
- Ask the learners to identify with talk partners different habitats around the world. Use a globe or a map of the world to find any particular places that they mention. Write a list of their responses. Discuss each habitat mentioned in turn. Think about the animals and plants that live there. Classify them as producers and consumers. Use any pictures that are available to support this discussion.

- Select the most appropriate photocopiable page to support the discussion and give it out to the learners. Ask them to complete the photocopiable page for that particular habitat.

Plenary

- Go through and discuss the range of responses from the learners to the photocopiable page given.
- Look at a globe or world map and decide where in the world such a habitat might be.

Ask the learners:

- Where in the world might this habitat be?
- Which producers could be at the start of the food chain in this habitat?
- Name a primary consumer from your food chain.
- What is the main predator in your food chain?
- Name two different animals that are prey in your food chain.

Support: Limit the number of plants and animals on each of photocopiable pages 97–104.

Extension: Ask these learners to construct food chains and to label them as fully as possible, using as many relevant terms as are necessary each time.

Name: _____

Food chains in the desert

1. Use the pictures at the bottom of the page to make a food chain that exists in the desert.

   Cut out the pictures and stick them in the correct order in the box below.

   Remember to use arrows!

2. Label the producer and consumers.

Name: _____

Food chains in a mountain environment

Here is a food chain from a mountain environment.

   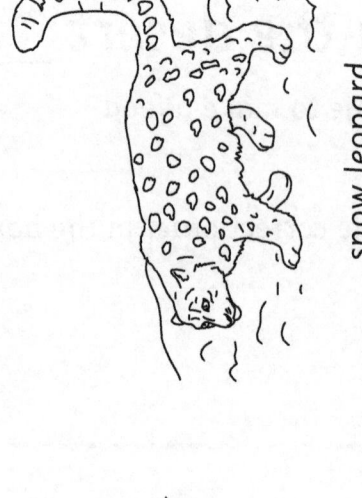 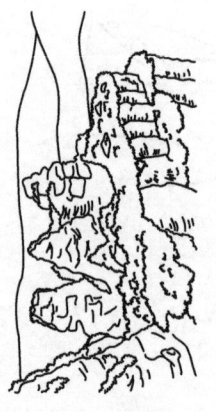

moss

mountain goat

snow leopard

1. Label the predator.

2. Complete these sentences.

a) The producer in this food chain is _____

b) The primary consumer is the _____

c) The predator is the _____

d) Its prey is the _____

Cambridge Primary: Ready to Go Lessons for Science Stage 6 © Hodder & Stoughton Ltd 2013

Food chains in the ocean

Here is a food chain from an ocean environment.

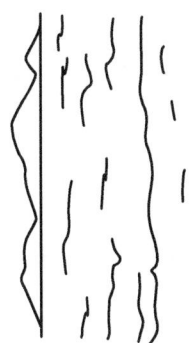

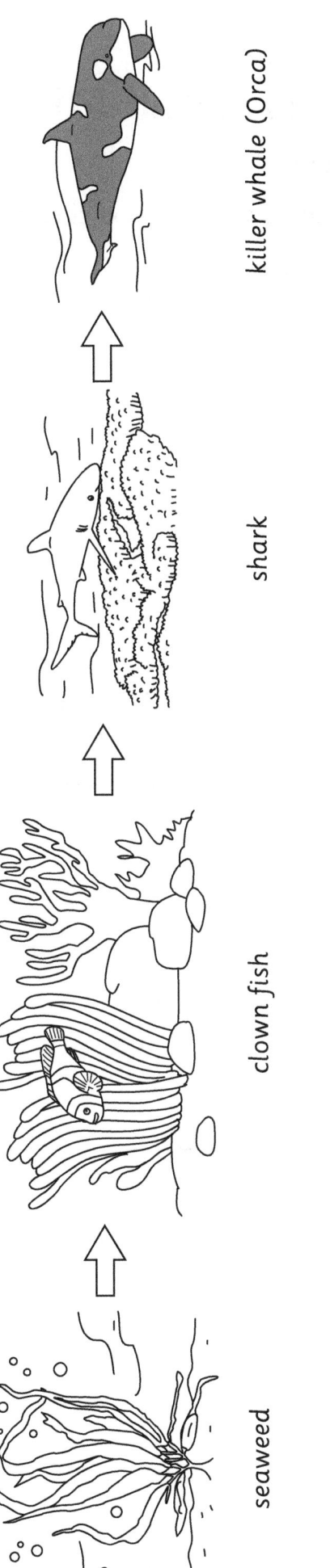

seaweed   clown fish   shark   killer whale (Orca)

Use the food chain to help you to answer these questions.

1. What is the name of the producer in this food chain? _____

2. Name the secondary consumer. _____

3. Which is the largest predator in this food chain? _____

4. What type of consumer is the clown fish? _____

5. Name a different producer found in the ocean. _____

Name: _____

A tree as a habitat

Trees provide habitats for many creatures.

Look at the pictures to help you write three different food chains for tree environments.

Use the internet or reference books to help you.

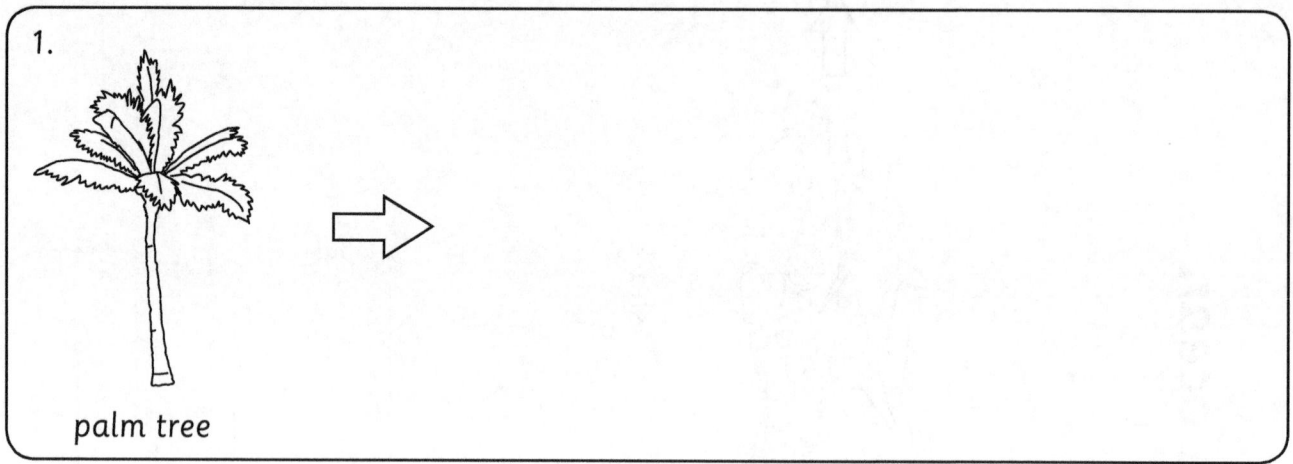

1.

palm tree

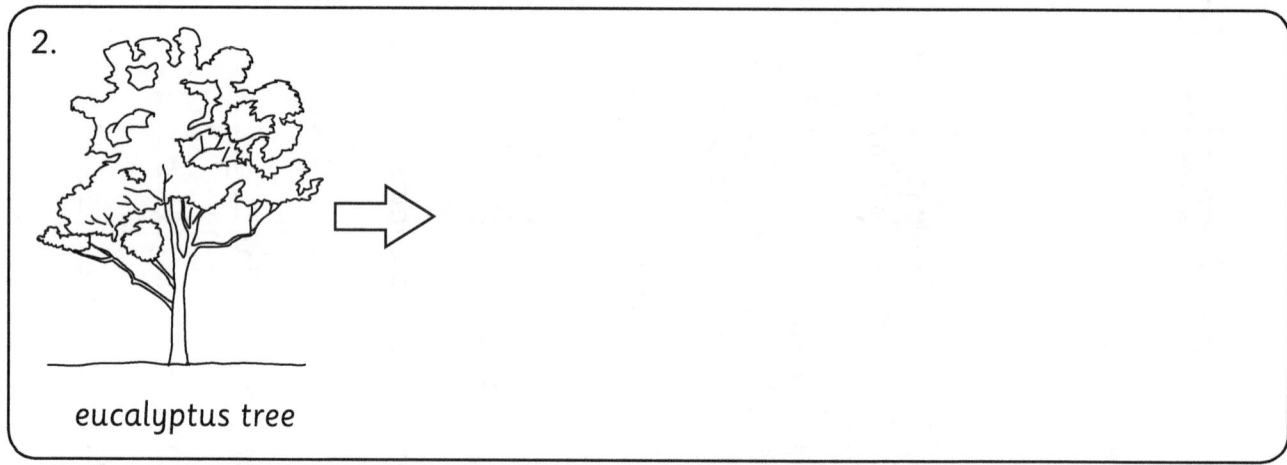

2.

eucalyptus tree

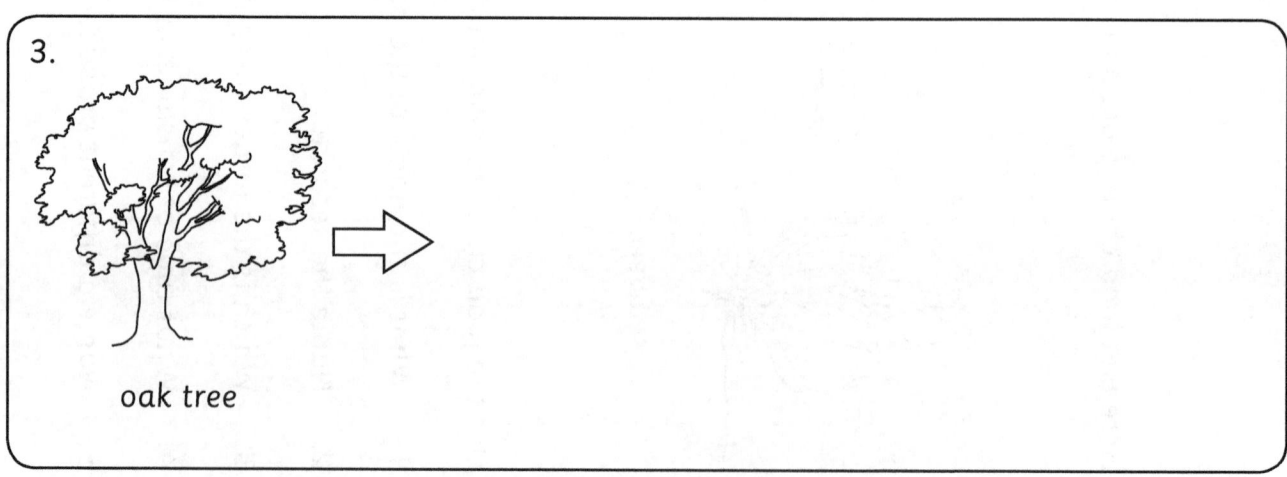

3.

oak tree

Cambridge Primary: Ready to Go Lessons for Science Stage 6 © Hodder & Stoughton Ltd 2013

Name: _____

The Antarctic

The Antarctic or Antarctica is a very cold region of the Earth near the South Pole.

(Globe diagram labelled: Europe, North America, Asia, Africa, South America, Antarctica)

1. Use reference pictures, books and the internet to make food chains using some of the animals and plants that are found in the Antarctic.
   Or use these pictures to help you instead.

plankton

krill

shrimp

seal

penguin

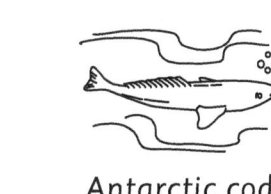

whale

albatross

Antarctic cod

2. How many different food chains can you make?

Name: _____

In the savannah plains

A savannah is an area that has lots of trees that are widely spaced, so sunlight can easily reach the ground.

Grass, small shrubs and thorn trees grow there.

Animals such as these all live there:

buffalo	cheetah	gazelle	giraffe	hyena	lion	zebra

Use this information to write three food chains for the savannah region.

Cambridge Primary: Ready to Go Lessons for Science Stage 6 © Hodder & Stoughton Ltd 2013

Name: _____

In the rainforest

1. Write a list of plants and animals that you know are found in a rainforest.

Plants	Animals

2. Use your list to draw or write a rainforest food chain.

Here are a few pictures to help you.

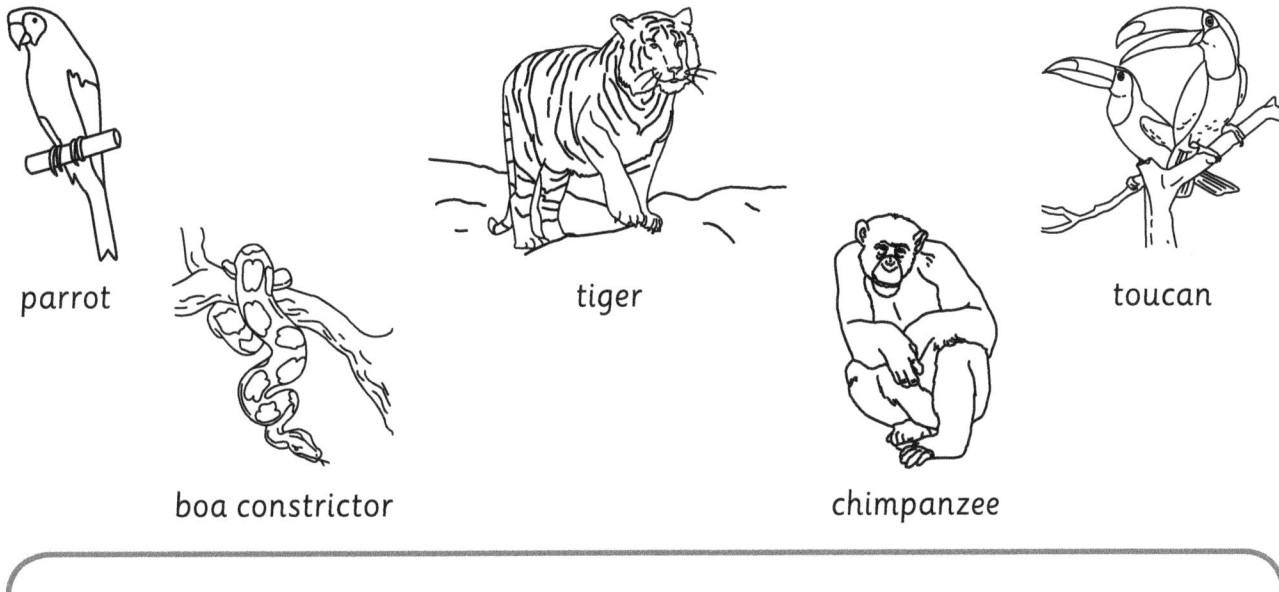

parrot

boa constrictor

tiger

chimpanzee

toucan

Name: _____

In a pond

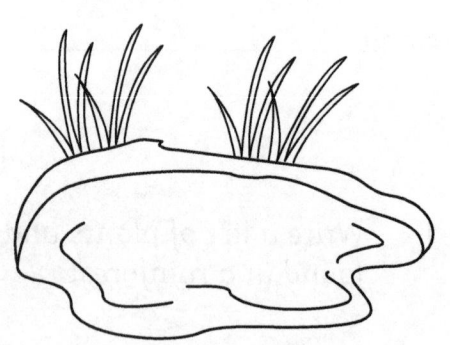

Here are some pond plants and animals.

Cut out the pictures and rearrange some of them to make two different food chains that could be found in a pond.

water lily | duckweed | newt | mosquito

goldfish | water beetle | frog | heron

Cambridge Primary: Ready to Go Lessons for Science Stage 6 © Hodder & Stoughton Ltd 2013

Unit assessment

Questions to ask

- What does a food chain represent or show? (The feeding relationships within a habitat.)
- What is a habitat?
- What is the ultimate source of energy for all food chains? (The Sun.)
- What do all food chains begin with? (A green plant or producer.)

- Describe the relationship between predator and prey. (The predator hunts and kills the prey for food.)
- What predators are there in a desert / mountain / pond / ocean / tree, and so on?

Summative assessment activities

Observe the learners while they participate in these activities. You will quickly be able to identify those who appear to be confident and those who may need additional support.

Making food chains

This activity assesses the learners' understanding of the structure of a food chain.

You will need:

A ball of string; scissors; a hole punch; pictures of animals and plants; sticky notes; felt-tipped pens or markers.

What to do

- Ask the learners working as individuals to use the pictures provided to create a food chain, joined together by string.
- Then ask them to use the sticky notes to add arrows to the chain and display it around the classroom.
- Observe their finished work.

Labelling food chains

This activity assesses the learners' understanding of key vocabulary.

You will need:

A set of labels containing the words: 'producer', 'primary', 'secondary', 'tertiary', 'consumer', 'predator', 'prey', 'herbivore', 'carnivore', 'omnivore'; some food chains in pictures and / or words with familiar and unfamiliar plants and animals in them.

What to do

- Give the labels to the learners and ask questions that require them to place the correct term against the correct animal or plant in the food chain.
- Differentiate the questions according to the level of the learners' ability.
- Record or photograph their completed activity.

Written assessment

Distribute photocopiable page 106. The learners should work independently, or with the usual adult support they receive in class.

Name: _____

Food chains assessment

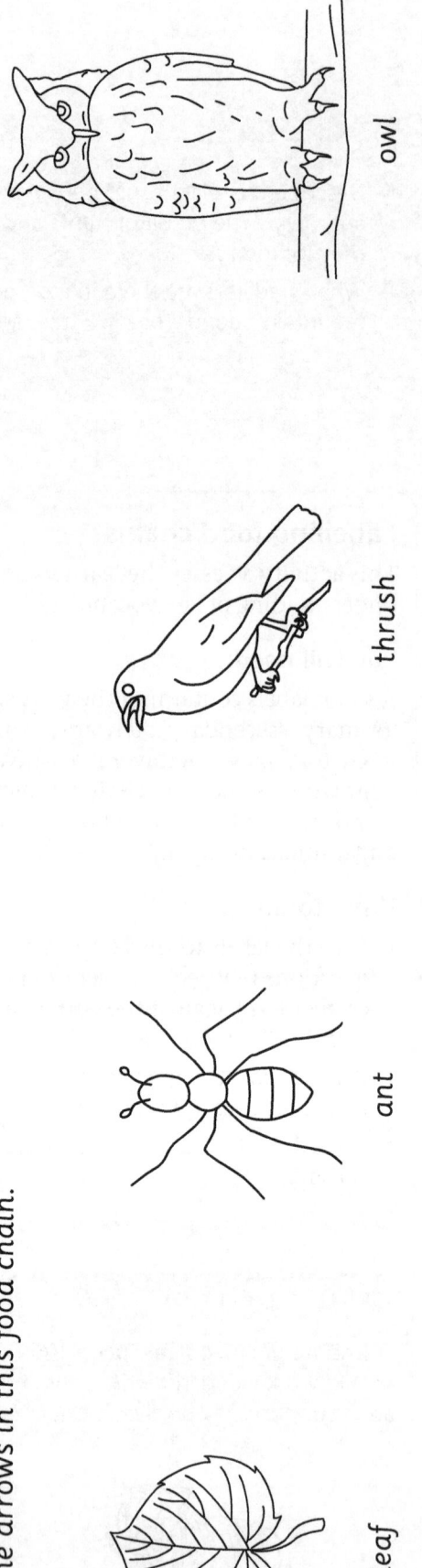

leaf   ant   thrush   owl

1. Put the arrows in this food chain.

2. Name the producer in this food chain. _____

3. What do we call the ant in this food chain? _____

4. How many predators are there in this food chain? _____

5. Write down one animal that is prey in this food chain. _____

6. What is the major predator for this animal? _____

7. If the secondary consumers became ill and died, how would this affect the other animals and plants in this food chain?
    _____
    _____

Cambridge Primary: Ready to Go Lessons for Science Stage 6 © Hodder & Stoughton Ltd 2013

Conducting electricity

Learning objectives

- Investigate how some materials are better conductors of electricity than others. (6Pm1)
- Make a variety of relevant observations and measurements using simple apparatus correctly. (6Eo1)
- Decide when observations and measurements need to be checked by repeating to give more reliable data. (6Eo2)

Resources

Circuitry equipment – cells (batteries), cell (battery) holders, lamps, wires, switches, crocodile clips, circuit boards; a range of materials for testing, e.g. wood, plastic, metal, paper, stone; photocopiable pages 108 and 109.

Starter

- Ask the learners in pairs or small groups to make a circuit containing one cell (battery), one lamp and one switch. Explain that they need to make the lamp light up and to control it with the switch. Do not tell them exactly what equipment they will need to do this – let them select what they need for themselves.
- Give out photocopiable page 108 for them to record their circuits on.
- Invite different pairs or groups to demonstrate their finished circuits. Compare similarities and differences between them, for example number of wires, position of the lamp and switch, and so on.

Main activities

- Ask one pair or group of learners if you can use their circuit. Remove the switch to break the circuit. Leave a gap where the switch has been. Ask the learners what you have done to the circuit (they need to have remembered the term 'break the circuit' from previous work on electricity). Remind them of this terminology.

- Explain that in this part of the lesson they will find out alternative materials to replace the switch and to make the circuit complete again. Use the phrase 'make the circuit', again as revision from previous learning.
- Either allow the learners free selection of materials to test in their circuits, or give them pre-selected sets of materials – depending on the availability of circuitry equipment and materials. Always make sure that there is plenty of spare circuitry equipment, especially lamps and cells (batteries).
- Give out photocopiable page 109 for the learners to record their results on.
- Allow them to carry out the test. Bring them back together as a whole class to discuss their findings.

Plenary

- Explain that any material that has completed the circuit did so because it is a good conductor of electricity. Something that conducts electricity allows electricity to flow through it.

Success criteria

Ask the learners:

- What do you need to do to make a working circuit?
- How does the switch work?
- Name a material that completed the circuit.
- Which materials did not complete the circuit?

Ideas for differentiation

Support: Organise these learners into mixed-ability groups, or pre-select familiar materials for them to test in their circuit.

Extension: Ask these learners to find three more materials or objects of their own choosing from around the classroom that conduct electricity.

Name: _____

Making circuits

1. List the exact amounts of equipment you used to make your circuit.

   - _____
   - _____
   - _____
   - _____
   - _____
   - _____

2. Draw your circuit below.

3. Comment on any changes you had to make to your circuit to make the lamp light up.

   _____

   _____

Cambridge Primary: Ready to Go Lessons for Science Stage 6 © Hodder & Stoughton Ltd 2013

Name: _____

Electrical conductors

1. Draw your circuit again, showing the gap where you removed the switch.

[   ]

2. Complete the table to show your results.

Material or object	Does it conduct electricity? (✓ or ✗)

Electrical insulators

● Investigate how some materials are better conductors of electricity than others. (6Pm1)

● Choose which equipment to use. (6Ep7)

● Use tables, bar charts and line graphs to present results. (6Eo3)

Resources

Circuitry equipment – cells (batteries), cell (battery) holders, lamps, wires, switches, crocodile clips, circuit boards; a range of materials for testing, e.g. wood, plastic, metal, paper, stone; pencils; pencils sharpened at both ends; photocopiable page 111.

Starter

• Before the lesson, set up a simple circuit – exactly the same as you asked the learners to make in the Starter activity in the previous lesson. Use one cell (battery), one lamp and one switch.

• Have available the same or a very similar set of materials and / or objects as in the previous lesson for the learners to choose from.

• Ask individual learners to come forward and choose an object or material (from the available selection) that is an electrical conductor. Ask them to insert this into the circuit in place of the switch. Ask them to predict what they think will happen before they insert the material.

• This will provide you with a good opportunity to observe how the learners adjust the circuit, if necessary, and hence gauge their understanding of making a complete circuit.

Main activities

• Ask the learners to think back to the previous lesson and to the materials that did not conduct electricity. Ask them if they know the scientific word for this group of materials. (Electrical insulators.)

• Explain that in this lesson they will find and identify more electrical insulators and conductors and group them accordingly.

• Give out photocopiable page 111 and ask the learners to record their results as they carry out the test.

• Organise the learners into pairs or small groups, depending on the availability of circuitry equipment and materials to carry out this activity.

• Allow them to do the activity as described on photocopiable page 111.

Plenary

• Bring the learners back together as a whole class to discuss their results.

• Hold up a variety of materials in turn and ask individual learners to identify them as electrical conductors or insulators. If there is any uncertainty, invite them to try out the material in the broken circuit.

Success criteria

Ask the learners:

● What is an electrical conductor?

● What do we call an object or material that does not conduct electricity?

● What can you do to find out if a material or object is an electrical conductor or insulator?

● Is a pencil an electrical insulator?

Ideas for differentiation

Support: Work with these learners in a small group, or organise the class into mixed-ability groups.

Extension: Ask these learners to test a pencil, then give them a pencil sharpened at both ends and ask them to test again.

Name: _____

Electrical insulators

Method (what to do)

- Construct a working circuit using a lamp, a switch and one cell (battery). (Think about what else you will need and collect all the equipment.)

- Remove the switch.

- Choose different materials to use to make the circuit.

- Record your results below.

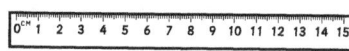

Results (what happened)

Insulators	Conductors

Conclusion (what you found out)

What kinds of materials are good electrical insulators?

_____

_____

Which materials are good conductors of electricity?

Learning objectives

- Investigate how some metals are good conductors of electricity while most other materials are not. (6Pm2)
- Make predictions using scientific knowledge and understanding. (6Ep4)
- Use results to draw conclusions and to make further predictions. (6Eo7)

Resources

Circuitry equipment – cells (batteries), cell (battery) holders, lamps, wires, switches, crocodile clips, circuit boards; a range of materials for testing, e.g. wood, plastic, a variety of different metals, paper, stone; flipchart and markers or whiteboard; photocopiable page 113.

Starter

- Write three names of materials on the flipchart or whiteboard, for example 'wood', 'paper' and 'gold'. Ask the learners: *Which is the odd one out and* **why**?
- Listen to their responses and praise good, scientific reasoning. This activity provides you with an opportunity to assess their knowledge of the properties of materials.
- Repeat this activity several times, using different examples of materials each time. However, make sure that the final list contains a metal or metals, for example silver, iron and copper. (Iron could be the odd one out because it rusts in air; copper could be the odd one out because it is not silver-coloured; silver could be the odd one out because it is used for precious jewellery, for example.)

Main activities

- Explain that in this lesson the learners will test materials in circuits again to identify them as conductors and insulators.
- Revise the terms **conductor** and **insulator** to make sure that all the learners know the difference.

- Before they begin the activity, ask each group to show you their working circuit. Organise the learners into pairs or groups according to the availability of resources and equipment for this activity. Alternatively, give each small group a similar set of different materials to test in their circuits and compare their findings at the end of the lesson.
- Give out photocopiable page 113 for them to record their findings and conclusion on.
- Ask the learners to look in particular at their list of conductors when they have completed the activity.

Plenary

- Invite the learners as individuals, pairs or small groups to feed their answers back to the rest of the class.
- Discuss any misunderstandings or misconceptions – emphasise that metals are good conductors of electricity.

Success criteria

Ask the learners:

- What does an electrical conductor do? (Allows electricity / electric current to pass / flow through it.)
- Give an example of an electrical conductor that you have used in this lesson.
- How could you group all the materials listed in your electrical conductors column of your results table? (Metals.)
- Do all metals conduct electricity?

Ideas for differentiation

Support: Give adult support to these learners or pre-select familiar materials for them to test.

Extension: Ask these learners to find one or more other metal objects around the classroom to test for electrical conductivity.

Name: _____

Good electrical conductors

Method (what to do)

- Construct a working circuit using a lamp, a switch and one cell (battery). (Think about what else you will need and collect all the equipment.)

- Remove the switch.

- Use the different materials you have been given to test the circuit.

- Record your results.

Results (what happened)

Only record the names of electrical conductors that you find.

Electrical conductors

Conclusion (what you found out)

What kinds of materials are good electrical conductors?

Electrical insulators in the home

Learning objectives

- Know why metals are used for cables and wires and why plastics are used to cover wires and as covers for plugs and switches. (6Pm3)
- Use tables, bar charts and line graphs to present results. (6Eo3)
- Make comparisons. (6Eo4)

Resources

Flipchart and markers or whiteboard; an electric plug with bared cable or wires and, if possible, a removeable cover; an electric iron; an electric kettle; an electric toaster or other kitchen gadget; photocopiable pages 115 and 116.

Starter

- Give out photocopiable page 115 to the learners in pairs or small groups. Explain that they will need to look around the classroom (and / or other designated areas) to find examples of electrical insulators being used in everyday life.
- Alternatively, divide the class into pairs or small groups and allocate them each a different area or classroom to investigate.
- Give the learners a set time limit (for example 15 minutes) to collect the observations and complete photocopiable page 115.

Main activities

- Gather all the learners back together to discuss and share their findings. (They should find wall light switches, plug sockets or extension leads, electrical cables, heater covers, and so on.)
- Write up a class list of all the examples found and display it prominently.
- Demonstrate the electrical plug and wiring / cable. Ask the learners to identify with talk partners the materials of the outer casing and inner components. (The outer casing will probably be rubber or plastic, the wires inside a plug are copper and the covering for cables is usually plastic.)

- Discuss why plastic is most often used as an insulator in these instances. (It is cheap to produce, it can be moulded easily into shape, it is waterproof, it doesn't react with air or water and it is widely available.)
- Give out photocopiable page 116 for the learners to complete. Explain that they have to write about some familiar everyday objects in terms of what the insulator is. Give them time to carry out the activity, then discuss their responses.

Plenary

- Make sure that the learners recognise copper as an orange-coloured metal.
- Discuss which materials are used to make the outer casing of small kitchen appliances – usually plastics.
- Ask the learners to think about good reasons for plastics being used in this way.

Success criteria

Ask the learners:

- What are the wires inside an electrical cable made from? (Copper.)
- Why is this a good metal to use? (It conducts electricity easily.)
- What insulating material is generally used for electrical plug and wall sockets? (Plastic, though some wall plates can be ornamental brass or metal.)
- Give as many reasons as you can think of why plastic is a good choice as an electrical insulator.

Ideas for differentiation

Support: Either work in a small group with these learners for the Starter activity, or arrange them to work in mixed-ability pairs or groups.

Extension: Ask these learners to find examples of insulators and conductors in their kitchen at home.

Name: _____

Electrical insulators around school

1. Find examples of electrical insulators in the classroom or another area with your partner or a small group. Complete the table below.

    An example has been done for you.

Item	Material	Reasons
kettle	plastic	waterproof can be moulded

2. Which materials are used a lot as electrical insulators?

    _____

    _____

Name: _____

Electrical insulators in the home

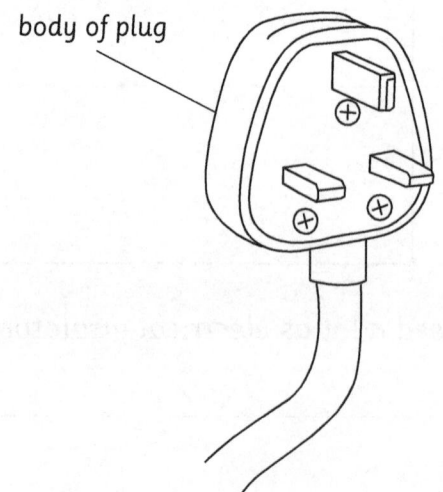

Saucepan

1. Which part of the saucepan is the insulator?

   _____

2. What material is this part of the saucepan made from?

   _____

Plugs

1. What are the prongs in the electrical plugs made from?

   _____

2. What is the body of the plug made of?

   _____

3. What are cables covered with?

   _____

4. Why?

   _____

   _____

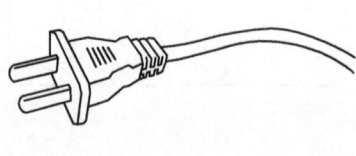

body of plug

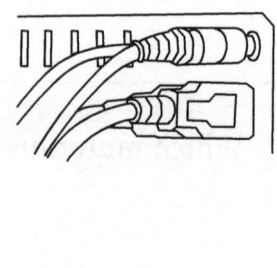

Cambridge Primary: Ready to Go Lessons for Science Stage 6 © Hodder & Stoughton Ltd 2013

Changing the brightness of lamps in a circuit

Learning objectives

- Predict and test the effects of making changes to circuits, including length or thickness of wire and the number and type of components. (6Pm4)
- Collect evidence and data to test ideas including predictions. (6Ep2)
- Say if and how evidence supports any prediction made. (6Eo9)

Resources

Circuitry equipment – cells (batteries), cell (battery) holders, lamps, wires, switches, crocodile clips, circuit boards; photocopiable pages 118 and 119.

Starter

- Set up a circuit for demonstration with lamps that glow brightly, for example use a 6V lamp powered by four 1.5V cells (batteries).
- Ask the learners to predict what will happen if you add another cell (battery) into the circuit.
- Explain that lamps are made to work with a specific amount of electricity (measured in volts) and that if this voltage is exceeded, the lamp will burn out. The same would apply if a motor was being used.
- On the basis of this explanation, it would be reasonable to expect the learners to predict that either the lamp(s) will glow brighter and / or burn out.

Main activities

- Explain that in this lesson the learners will investigate what affects the brightness of lamps in a circuit.
- Ask the learners to discuss with talk partners what they might like to try. (They might suggest adding more or taking some lamps out of the circuit. They might also suggest using a cell [battery] with more or less voltage.)
- Give out photocopiable pages 118 and 119 for them to write their predictions and draw circuits on.

- Organise the class into pairs or small groups, according to the amount of equipment and resources available. Ask each pair or group to make a working circuit that you check before they continue with their investigation. Alternatively, you could direct one group to test the effect of using fewer or more lamps and other groups to test the effects of using fewer or more cells (batteries).

Plenary

- Set up different circuits with different numbers of cells (batteries) or lamps in them, or use the circuits made by the learners in the course of their investigations.
- Ask them to predict the relative brightness of the lamps in the different circuits, with reasons why.
- Explain that if there are more cells (batteries) in the circuit, the circuit is doing more work – which makes the lamps glow more brightly. If too many cells (batteries) are used, the lamp(s) will blow and not light up any more.

Success criteria

Ask the learners:

- What is the best combination of lamps and cells (batteries) to make the lamps glow brightest?
- How many is too many lamps for the circuit?
- How can you tell?
- How can you decide how many cells (batteries) and lamps to use without blowing the circuit?

Ideas for differentiation

Support: Work with these learners in a small group and select the task for them to investigate – either the number of lamps or the number of cells (batteries).

Extension: Challenge these learners to find the combination of cells (batteries) and lamps in a circuit which makes the lamps glow brightest.

Name: _____

Changing the brightness of lamps in a circuit 1

Prediction

1. What might affect the brightness of lamps in a circuit?

   List as many things as you can think of.

   _____

   _____

   _____

   _____

2. Draw the circuit you will use to carry out your investigation in the box below.

   Include these components:

   > lamp(s)   cell(s) (battery/ies)   wires

Cambridge Primary: Ready to Go Lessons for Science Stage 6 © Hodder & Stoughton Ltd 2013

Name: _____

Changing the brightness of lamps in a circuit 2

3. Which factor will you change each time?

_____

Results (what happened)

4. Draw the different circuits that you tried.

5. What makes the lamps brighter?

_____

6. Was your prediction correct? yes / no

Changing the number of cells (batteries) in a circuit

Learning objectives

● Predict and test the effects of making changes to circuits, including length or thickness of wire and the number and type of components. (6Pm4)

● Collect evidence and data to test ideas including predictions. (6Ep2)

● Say if and how evidence supports any prediction made. (6Eo9)

Resources

Circuitry equipment – cells (batteries), cell (battery) holders, lamps, wires, switches, crocodile clips, circuit boards, buzzers and motors; photocopiable pages 121 and 122.

Starter

• Ask the learners to discuss with talk partners where the power inside a cell (battery) comes from. (The chemicals inside the cell [battery].)

• Ask the learners what happens when you add another cell (battery) into a circuit. If they have good understanding from the previous lesson, they should be able to tell you that it will supply more electricity to the circuit. This means that, for example, a lamp would glow more brightly or a motor in the circuit would spin faster.

• Demonstrate the difference by showing two circuits – one with one cell (battery) and one lamp, the other with two cells (batteries) and one lamp. Discuss the difference in brightness between the two lamps. (The circuit with one cell [battery] and one lamp will be quite bright; the circuit with two cells [batteries] and one lamp will be brighter.)

Main activities

• Explain that in this lesson the learners will make and compare circuits with different numbers of components in them. They will predict first, then make and compare the circuits. Make sure that the learners understand what the components are – cells (batteries), lamps, wires, motors, switches, and so on.

• Give out photocopiable pages 121 and 122, which ask the learners to make various circuits and compare them for the brightness of the lamps in them. Ask the learners to look at the pictures of the circuits on photocopiable page 121 and predict which will be the brightest. Look at their responses before allowing them to do the activities. Photocopiable page 122 requires the learners to insert a buzzer or a motor into a circuit and to observe the effects.

• Organise the class into pairs or small groups and decide if they should work in ability or mixed-ability groups. This could depend on the time available and also the equipment and other resources available.

• Ask the learners to carry out the activities and complete photocopiable pages 121 and 122.

Plenary

• Bring the learners back together and discuss their answers to photocopiable pages 121 and 122.

• Explain that the more cells (batteries) there are in a circuit, the brighter the lamp will be.

• Discuss their predictions.

Success criteria

Ask the learners:

● If you add more cells (batteries) to a circuit, how does this affect the brightness of the lamp(s)?

● What do more cells (batteries) supply the circuit with? (More power / electricity.)

● How would a motor change if more cells (batteries) are added to a circuit?

● What effect would adding more cells (batteries) have on a buzzer in a circuit?

Ideas for differentiation

Support: Work with these learners in a small group or put them into mixed-ability groups to carry out the activities.

Extension: Ask these learners what effect changing the number of cells (batteries) would have on a buzzer in a circuit.

Name: _____

Changing the number
of cells (batteries)

Predict

1. For each pair of circuits, tick (✓) the one that will have the brightest lamp.

a)

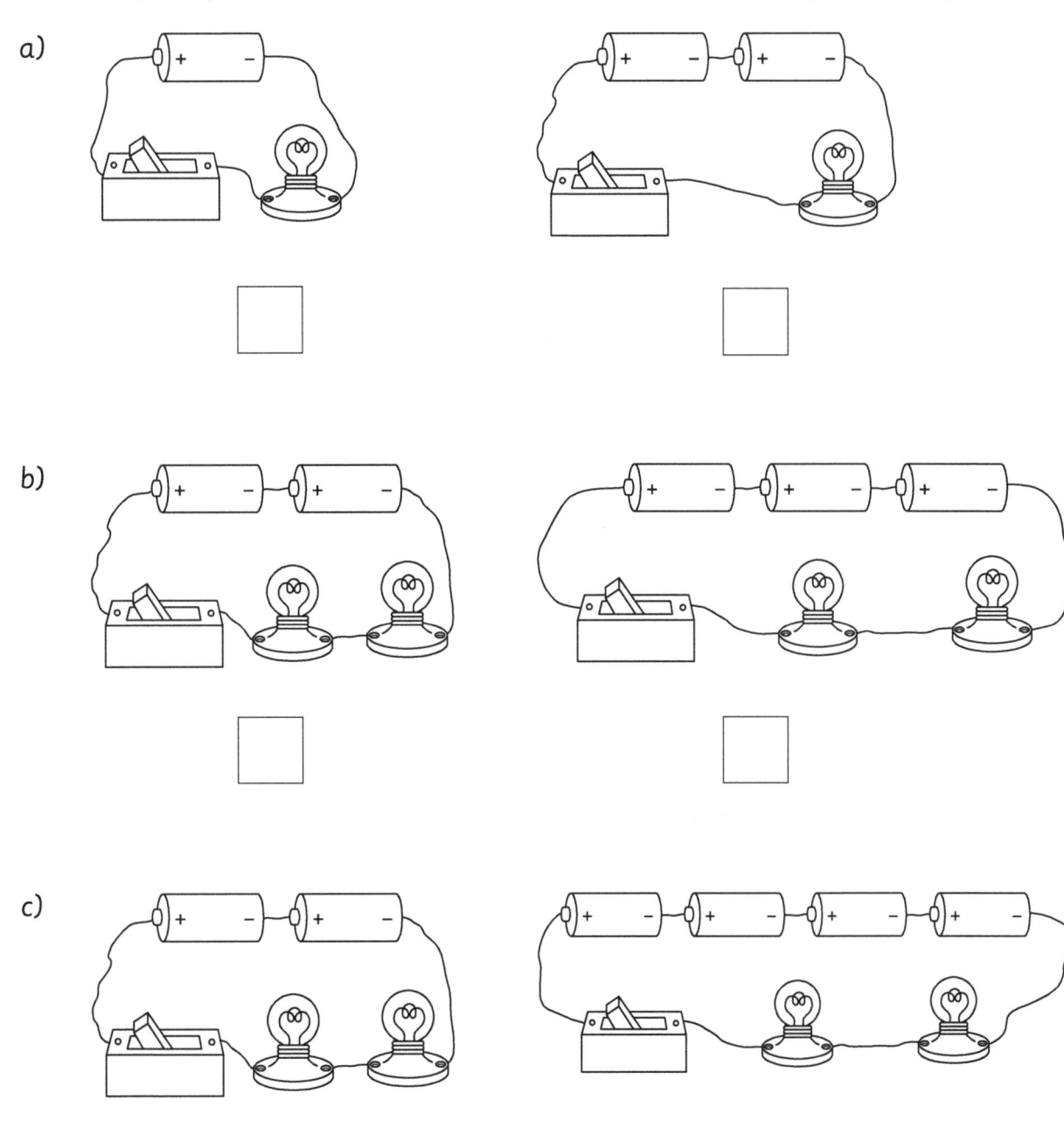

b)

c)

2. Now make each pair of circuits to test your predictions.

3. How many of your predictions were correct? _____

Name: _____

Improving a circuit

1. Set up a circuit with a motor instead of lamp(s).

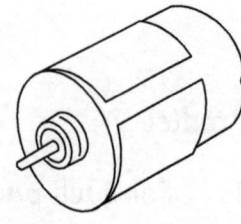

2. Draw a circuit that makes the motor spin faster.

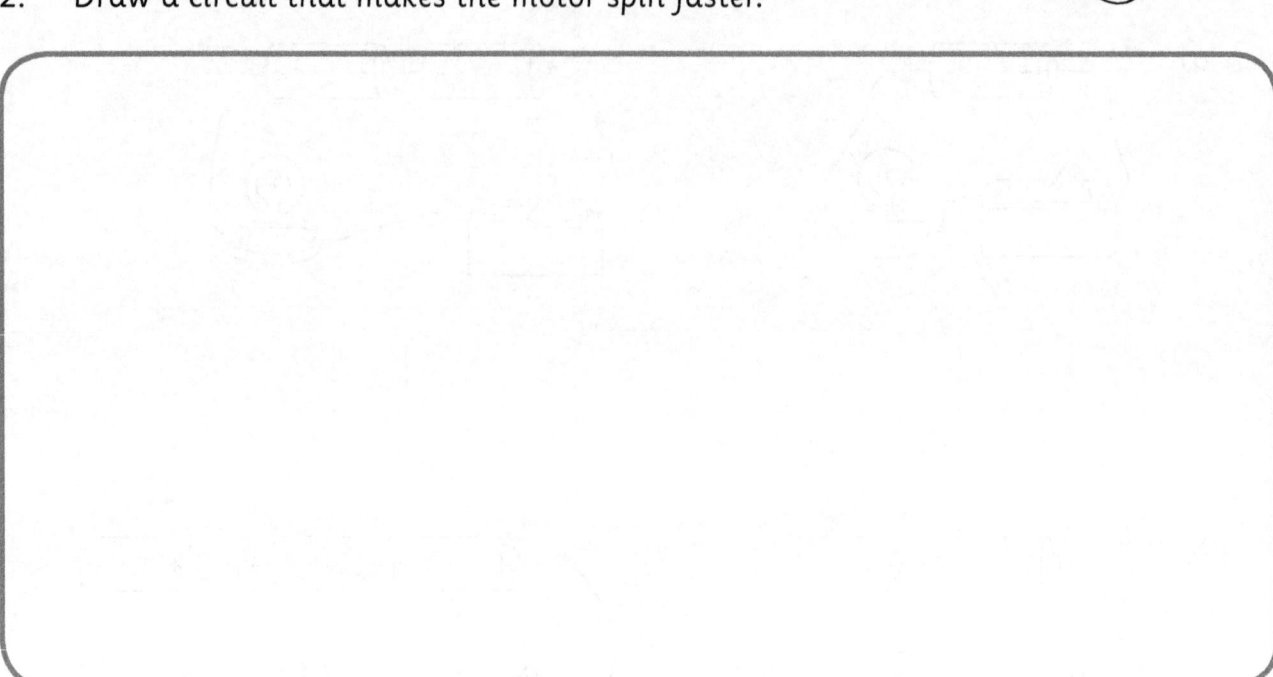

3. Now set up a circuit with a buzzer instead of lamp(s).

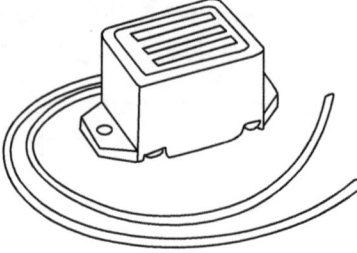

4. Draw a circuit that makes the buzzer sound loudly.

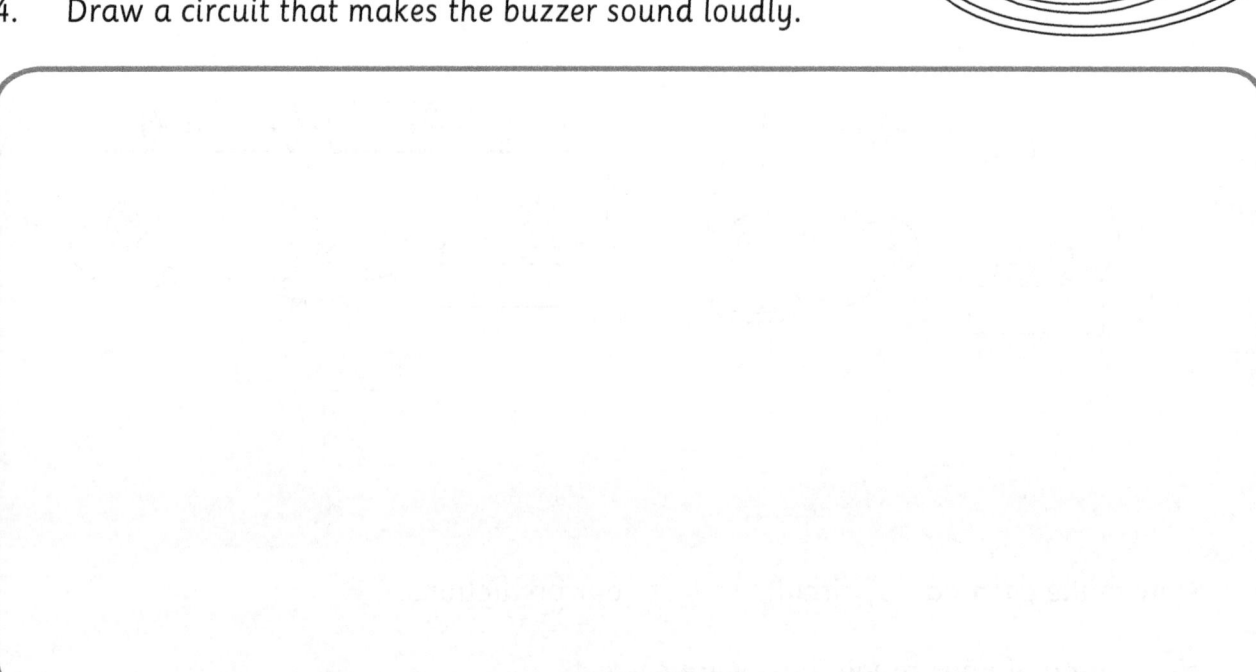

 Cambridge Primary: Ready to Go Lessons for Science Stage 6 © Hodder & Stoughton Ltd 2013

Changing simple circuits

- Predict and test the effects of making changes to circuits, including length or thickness of wire and the number and type of components. (6Pm4)
- Identify patterns in results and results that do not appear to fit the pattern. (6Eo6)
- Say if and how evidence supports any prediction made. (6Eo9)

Resources

Circuitry equipment – cells (batteries), cell (battery) holders, lamps, wires, switches, crocodile clips, circuit boards; photocopiable pages 124 and 125.

Starter

- Ask the learners in pairs to make a working circuit using one cell (battery), a switch, a lamp and wires.
- Ask the learners to join up with another pair. Ask them to predict what would happen if they added in another cell (battery). Listen to their predictions. Ask them, in their groups of four, to keep one of the original circuits and put an extra cell (battery) in the other circuit. They should then compare the brightness of the lamp in each circuit. (Make sure that the cells [batteries] and lamps are of the correct voltage.)

Main activities

- Introduce the term **series**. Explain that this means a circuit in which each component is connected to the next in a line, in turn, just as a TV series is made up of one episode after another and books written in series contain the story written in order, one after the other.
- Explain that the purpose of this lesson is for them to try out and observe what happens when you add more cells (batteries) or more lamps in a series circuit.

- Give out photocopiable pages 124 and 125 for the learners to follow the instructions, then make and observe the circuits. Organise the learners into pairs or small groups or have half the class testing different numbers of cells (batteries) and the other half of the class testing the number of lamps.
- Circulate the room, noting how well or badly they are able to follow the instructions and carry out the activities. Offer advice, support and help as necessary.

Plenary

- Ask the learners as a class to demonstrate the different circuits they have made and to compare the differences.
- Make sure that the learners recognise that if you add more cells (batteries) in a line to a circuit, the lamp will be brighter.
- Explain that when more lamps are added to a circuit in a line, the lamps glow more dimly.

Success criteria

Ask the learners:

- What does 'in series' mean?
- When more cells (batteries) are added to a circuit in series, what happens to the brightness of the lamp?
- When more lamps are added to a circuit in series, what happens to the brightness of the lamp(s)?

Ideas for differentiation

Support: Work with these learners in a small group or organise them into mixed-ability groups for this activity.

Extension: Ask these learners to predict and test what happens when more cells (batteries) and lamps are added into a series circuit.

Name: _____

Changing simple circuits 1

1. Make a circuit using one cell (battery),
    one switch, one lamp and wires.
    Draw your circuit in the box below.

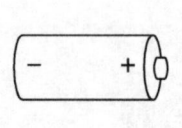 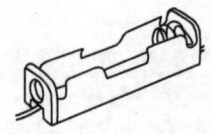 

 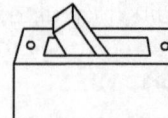 

2. Now add in another cell (battery). Draw this circuit in the box below.

3. How does the brightness of the lamp compare in each circuit?

    a)   In circuit 1 the lamp is _____.

    b)   In circuit 2 the lamp is _____.

Name: _____

Changing simple circuits 2

1. Make a circuit using one cell (battery), one switch, two lamps and wires. Draw your circuit in the box below.

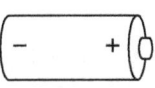   

 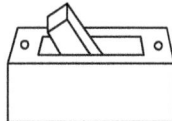 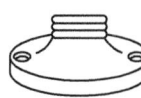 

2. Now add in another lamp. Draw this circuit in the box below.

Conclusion (what you found out)

3. How does the brightness of the lamp compare in each circuit? Complete this sentence.

    The _____ the number of lamps, the _____ the light from the lamps.

4. Is the lamp nearest to the cell (battery) the brightest? yes / no

Testing the thickness of wires in a circuit

Learning objectives

- Predict and test the effects of making changes to circuits, including length or thickness of wire and the number and type of components. (6Pm4)
- Identify patterns in results and results that do not appear to fit the pattern. (6Eo6)
- Suggest and evaluate explanations for predictions using scientific knowledge and understanding and communicate these clearly to others. (6Eo8)

Resources

Different types of wire – mains cable, fuse wire (different thicknesses), circuit wire / plastic-coated wire; photocopiable page 127; glue or staples and stapler; scissors; photocopiable page 128; circuitry equipment – cells (batteries), cell (battery) holders, lamps, wires, switches, crocodile clips, circuit boards.

Starter

- Show the learners the different types of wires available. Discuss where each type of wire might be used, for example in a plug, for a TV or in a computer cable. Remind the learners of the dangers of playing with mains electricity.
- In pairs or small groups, give the learners a selection of different thicknesses of wire – suitable for testing in a circuit.
- Ask them to arrange the wires in order from thinnest to thickest. Give out photocopiable page 127 for them to stick samples of the wires on to – use staples or glue for this. Ask them to predict which will give the brightest lamp.

Main activities

- Ask the learners: *How does the thickness of the wire affect the brightness of a lamp in a circuit?*
- Listen to the learners' responses, but do not comment on them. Explain that in this lesson they will need to plan and carry out a fair test to answer this question.

- Allow them to choose their working groups – these could even be friendship groups. Alternatively, choose the working groups for the learners.
- Approve their plans on photocopiable page 127 before they begin their testing, and check that their circuit is working before allowing them to commence work.
- Allow the learners to carry out the investigation and to collect their evidence. Discuss each group's findings as you visit each group in turn as they work.
- Give out photocopiable page 128 now for them to record their method, results and conclusions on.

Plenary

- Invite each group in turn to report their findings to the rest of the class.
- Discuss any similarities and / or differences in their results.

Success criteria

Ask the learners:

- Which wire produced the brightest light from the lamp in the circuit?
- Which wire made the lamp glow dimly?
- How did you make it a fair test?
- How does the thickness of the wire affect the brightness of a lamp in a circuit?

Ideas for differentiation

Support: Give these learners fewer pieces of wire to test.

Extension: Challenge these learners to find out which length of the best wire produces the brightest light from the lamp in the circuit.

Name: _____

Testing the thickness of wires in a circuit 1

Prediction

1. Stick the samples of wire you will use in the box below.
    Arrange them in order from thinnest to thickest. Number each wire.

Thinnest

Thickest

2. I predict that the wire that will make the lamp glow brightest in a circuit will

    be wire number _____ .

3. Draw your circuit for testing in the box below.

Name: _____

Testing the thickness
of wires in a circuit 2

Method (what you did)

4. Write about what you did.

_____

_____

_____

_____

_____

Results (what happened)

Brightest = wire number _____ .

Dimmest = wire number _____ .

Conclusion (what you found out)

5. Which wire produced the brightest light? _____

6. Why? _____

_____

7. Complete the sentence below:

The _____ the wire, the _____ the lamp.

Cambridge Primary: Ready to Go Lessons for Science Stage 6 © Hodder & Stoughton Ltd 2013

Dimmer switches

● Predict and test the effects of making changes to circuits, including length or thickness of wire and the number and type of components. (6Pm4)

● Collect evidence and data to test ideas including predictions. (6Ep2)

● Suggest and evaluate explanations for predictions using scientific knowledge and understanding and communicate these clearly to others. (6Eo8)

Resources

Circuitry equipment – cells (batteries), cell (battery) holders, lamps, wires, switches, crocodile clips, circuit boards; internet access or a dimmer switch; lengths of wire marked in 1 cm intervals with a marker pen; photocopiable pages 130 and 131.

Starter

• Show some actual circuits, or pictures of circuits, containing several components. Remind the learners that the more components there are in a circuit, the smaller the current will be flowing through it, so the lamp will be dimmer or the motor will be slower or the buzzer will not be so loud.

• Ask the learners to think back to the previous lesson: *Which wire made the lamp glow brightest?* (The thicker wire.)

• Ask if any of the learners have dimmer switches in their homes. Talk about visits to the cinema when the lights are dimmed before a performance and explain that this is an example of a dimmer switch being used.

• Demonstrate how a dimmer switch works by showing the learners the inner workings (it has a coil of wire inside). Alternatively find a picture of one on the internet. Explain that as the switch is turned, it alters the length of wire in the coil and this alters the brightness of the light produced.

Main activities

• Explain that in this lesson the learners will investigate how the length of a wire affects the brightness of a lamp in a circuit.

• Set up and demonstrate what the circuit should look like in order for them to do the test (but do not actually connect the crocodile clip to the wire).

• Give out photocopiable pages 130 and 131 for them to follow to carry out the investigation. Talk through the instructions on photocopiable page 130 and use this to show the circuit they will need. Answer any questions that arise.

• Organise the class into working groups, as appropriate – ability or mixed-ability groups.

• Allow them to carry out the investigation and to record their results.

Plenary

• Discuss the learners' results and compare their findings.

• Explain that the longer the wire, the smaller the flow of current and so the dimmer the lamp.

Success criteria

Ask the learners:

● Which length of wire produced the brightest light?

● How long was the wire when the lamp was dimmest?

● What was the factor that you changed each time?

● How did you make this a fair test?

Ideas for differentiation

Support: Provide these learners with adult support to supervise the accuracy of their measurements.

Extension: Challenge these learners to see if a different thickness of wire produces similar or different results at the same lengths.

Making a dimmer switch 1

1. Construct a circuit like the one shown here.

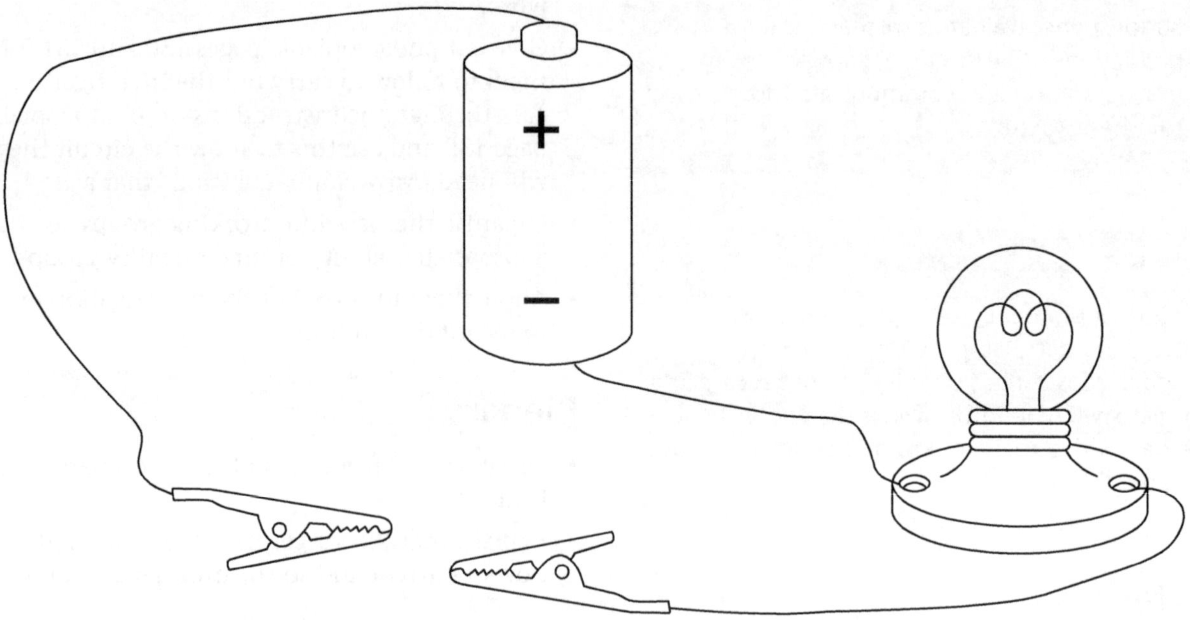

You will need:

A length of wire marked off in cm divisions, a ruler.

Method (what to do)

- Use the crocodile clips to place the wire in the circuit.

- Move the crocodile clips further apart or closer together so that there is a different length of wire between them each time.

- Look at the brightness of the lamp each time.

- Record your results and conclusion on the next page ('Making a dimmer switch 2').

Cambridge Primary: Ready to Go Lessons for Science Stage 6 © Hodder & Stoughton Ltd 2013

Name: _____

Making a dimmer switch 2

Results (what happened)

2. Draw or write about what happened in the box below.

Conclusion (what you found out)

3. a) Which length of wire gave the brightest light? _____

    b) Explain why.

    _____

    _____

    _____

Circuit symbols

Learning objectives

- Represent series circuits with drawings and conventional symbols. (6Pm5)
- Make comparisons. (6Eo4)

Resources

Circuitry equipment – cells (batteries), cell (battery) holders, lamps, wires, switches, crocodile clips, circuit boards; whiteboard or flipchart and markers; a set of laminated circuit-symbol cards from photocopiable page 133; torches and cells (batteries); photocopiable pages 134 and 135; sharp pencils; rulers; internet access.

Starter

- Prepare a circuit with one cell (battery), one lamp, a switch and wires.
- Invite a learner to come forward and draw a picture of this circuit.
- Praise their efforts and discuss the time it takes to draw such a picture.
- If there is time, repeat the activity, asking other learners to draw pictures of circuits. Alternatively, ask some of the learners to make circuits for a partner to draw pictures of.
- Retain these pictures for reference during the Main part of the lesson.

Main activities

- Explain that over time, scientists have developed a universally acceptable system of circuit symbols. These include simple representations of the components of a circuit, which can be drawn quickly and easily and save a lot of time.
- Pre-prepare a set of circuit-symbol cards from photocopiable page 133, laminated if possible. Show each card and discuss the symbol in turn. Alternatively, cover up the label, show the learners the symbol and ask them to guess what the symbol might represent, or produce separate cards of symbols and words and ask the learners to match them.
- Demonstrate how to draw the circuit diagram for the circuit you began the Starter activity

with. If possible, draw the circuit diagram alongside the circuit picture from the Starter activity. It is useful for the learners to see the picture and diagram side by side to see how the circuit symbols relate to the picture. Demonstrate drawing straight lines with a ruler and not leaving any gaps between wires and components (apart from in between the terminals of the cell [battery] in the symbols for cells [batteries]. Make sure to use only the symbols as drawn on photocopiable page 134.

- Give out photocopiable page 134, which contains the circuit symbols to be used. Go through the page with the learners.
- Organise the learners into pairs or small groups and ask them to construct, then use the circuit symbols to draw, a circuit diagram for each circuit they make on photocopiable page 135.

Plenary

- Invite different learners to show the rest of the class the circuits they have drawn on photocopiable page 135.
- Play the interactive game at www.scibermonkey.org: Go to 7–11 → Energy and electricity → Electrical circuits and conductors → Circuit experiments. This activity allows the learners to apply their learning of interpreting circuit diagrams and also revises adding or removing components in circuits.

Success criteria

Ask the learners:

- Why do we use circuit symbols?
- What is this circuit symbol for? (Show a circuit-symbol card, covering up the label.)
- Draw the symbol for a … (Use individual whiteboards for this, if available.)

Ideas for differentiation

Support: Work with these learners in a small group.

Extension: Ask these learners to draw a circuit picture and diagram for how a torch works.

Circuit symbol cards

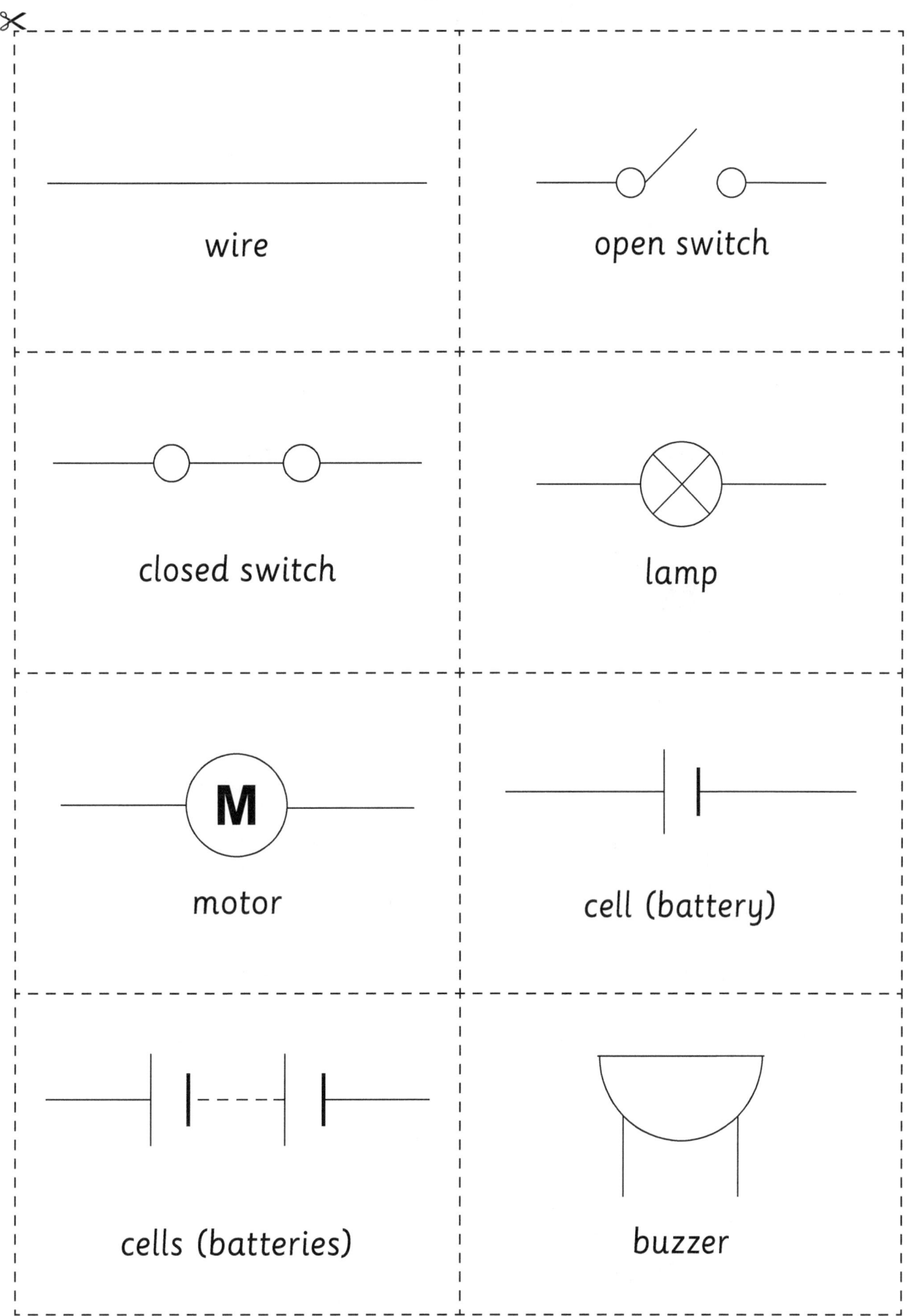

wire

open switch

closed switch

lamp

motor

cell (battery)

cells (batteries)

buzzer

Circuit symbols

We use these symbols to draw diagrams of circuits:

wire

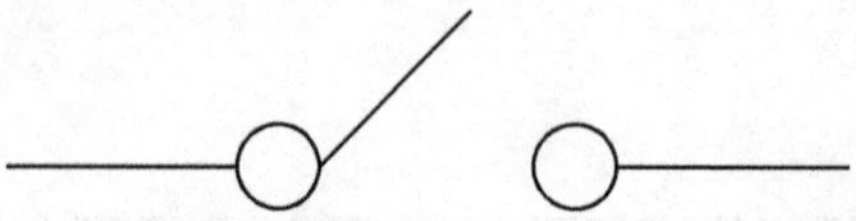

open switch

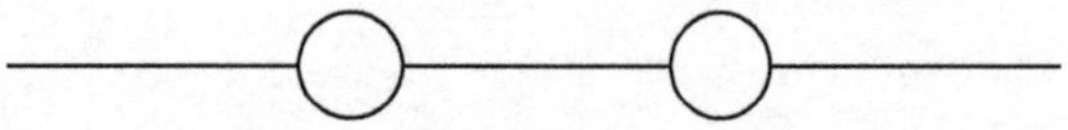

closed switch

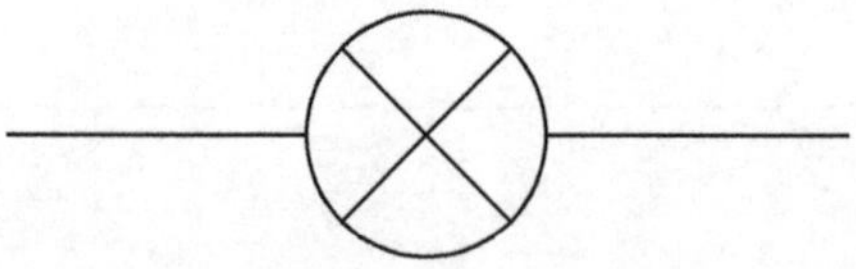

lamp

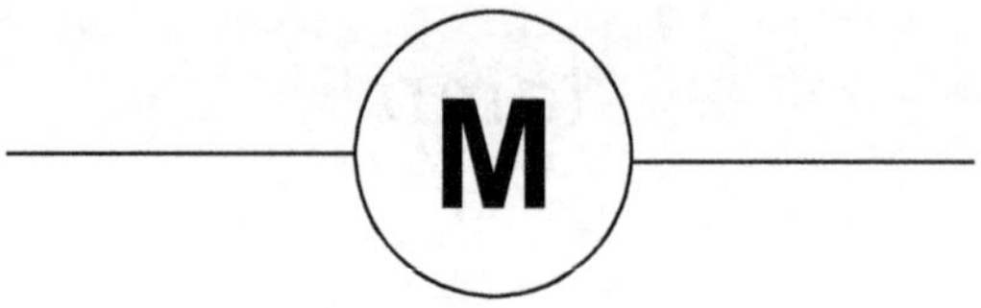

motor

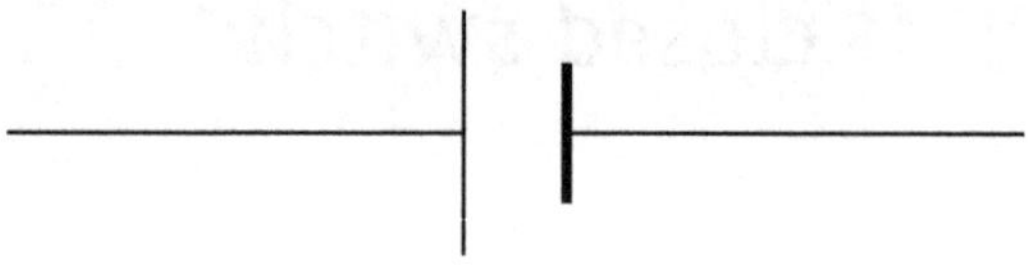

cell (battery)

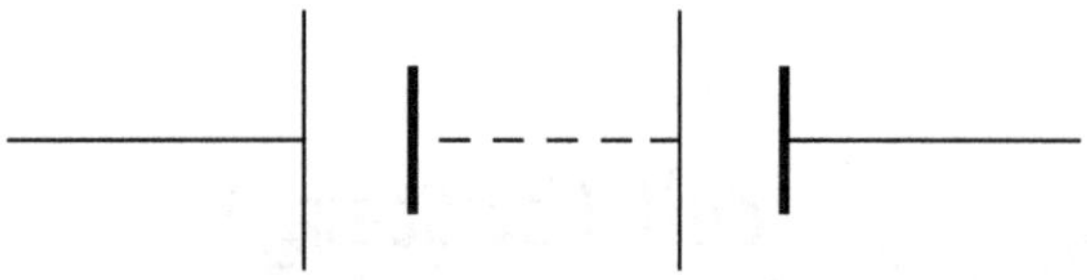

cells (batteries)

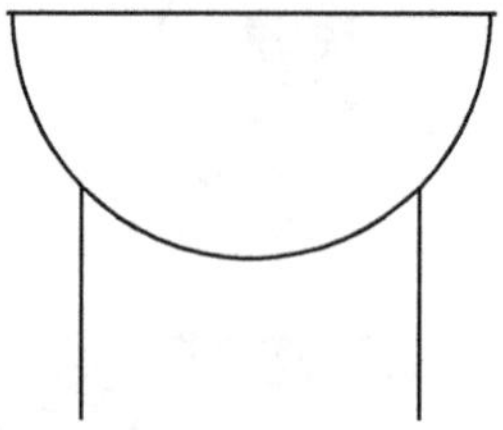

buzzer

Circuit diagrams provide a quick, easy way to draw circuits, for example:

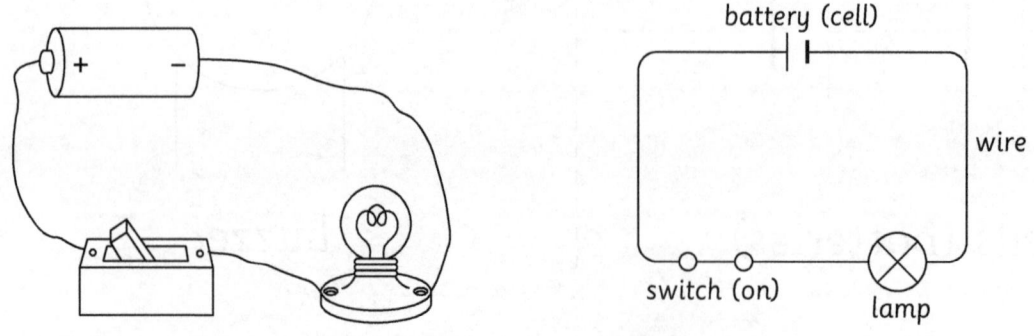

battery (cell)

wire

switch (on)

lamp

Cambridge Primary: Ready to Go Lessons for Science Stage 6 © Hodder & Stoughton Ltd 2013

Name: _____

Drawing circuits

1. Make a circuit.

2. Draw a picture of it in the box below.

3. Now use the symbols on the 'Circuit symbols' page to draw a diagram of the circuit you have made.

Unit assessment

- What materials are good electrical conductors?
- What are materials called that do not conduct electricity?
- Why is plastic used as a covering for electrical cables?

- What effect does changing the number of lamps have on a circuit?
- If you add more cells (batteries) to a circuit, what happens?
- What is the circuit symbol for a lamp?

Summative assessment activities

Observe the learners while they participate in these activities. You will quickly be able to identify those who appear to be confident and those who may need additional support.

Conductors and insulators

This activity assesses the learners' ability to classify objects as conductors or insulators.

You will need:

An empty drinks can, a candle, a flower, a key, a metal fork, a plastic spoon, a rock, a tooth or bone, a twig, a pair of scissors (alternatively, provide pictures of each of these items); recording pages: prepare an A4 page with two columns, headed 'Conductors' and 'Insulators' if you prefer to do this as a written rather than oral response.

What to do

- Ask the learners as individuals, pairs or small groups (depending on time and equipment available), to sort the objects / pictures into conductors and insulators. (If you are doing the written activity, ask the learners to write their responses on the recording page.)
- Differentiate the activity by asking learners of different abilities to classify more or fewer objects.
- Record the learners' scores on a class checklist, or make written notes as necessary.

Making circuits

This activity assesses the learners' ability to make a circuit.

You will need:

A range of circuitry equipment – lamps, cells (batteries), wires, motors, buzzers, and so on; digital camera; paper.

What to do

- Ask learners individually to make a circuit containing a particular combination of components – either pre-select the components to be used, or allow them to choose from the complete range available.
- Ask them to predict what would happen if they had more or fewer lamps or cells (batteries) (specify the actual number).
- Then allow them to remove or insert the extra components to check their prediction.
- Record their predictions and the validity of such – or ask the learners to draw and write their own predictions and to draw or take photographs of the circuits they have constructed.

Distribute photocopiable pages 137, 138 and 139. The learners should work independently, or with the usual adult support they receive in class.

Name: _____

Using circuits

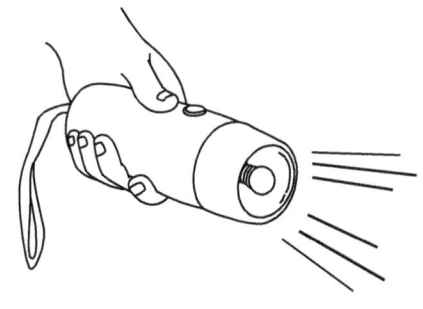

Pablo has a new torch.

He wants to make it shine more brightly.

1. What could he do?

    _____

    _____

2. If you have a circuit with one lamp and connect another cell (battery) into it, what happens to the lamp? Tick (✓) one box.

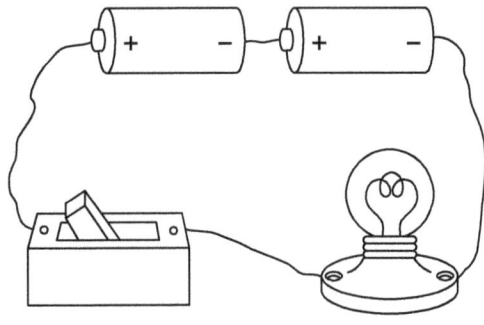

    a)   The lamp will glow brighter.   ☐

    b)   The lamp will glow the same.   ☐

    c)   The lamp will be dimmer.   ☐

3. If a lamp is labelled 3V (three Volts) and you connect it to a 9V (nine Volt) cell (battery), what will happen?

    _____

4. List some jobs for which a torch might be used.

    _____

    _____

Name: _____

How do circuits work?

Anna has a new toy car.

It is cell (battery) operated.

1. What could she do to make the car go faster?

    _____

    _____

2. Draw a circuit diagram of a series circuit containing two cells (batteries), two lamps and a switch.

3. What would happen if the cells (batteries) were not inserted (put in) in the correct way?

    _____

    _____

 Cambridge Primary: Ready to Go Lessons for Science Stage 6 © Hodder & Stoughton Ltd 2013

Name: _____

Circuit symbols

1. Complete the table by drawing the correct circuit symbols.

Equipment	Circuit symbol
cell (battery)	
lamp	
buzzer	
motor	
switch	
wires	

2. Why does a circuit have to be complete?

_____

_____

How humans can affect the environment

Learning objectives

- Explore how humans have positive and negative effects on the environment, e.g. loss of species, protection of habitats. (6Be1)
- Identify factors that are relevant to a particular situation. (6Ep6)
- Make comparisons. (6Eo4)

Resources

Flipchart and markers or interactive whiteboard; photocopiable page 141.

Starter

- Ask the learners what is meant by the term **environment**. Discuss their answers and agree a joint definition. Write and display this prominently. A simple definition of **environment** is 'the surroundings or conditions in which humans, animals or plants live or survive'. Your school environment is the buildings and grounds – the whole school site.
- Discuss ways in which humans care for the environment. Listen to the learners' suggestions – lead the discussion to cover such points as keeping it clean, making sure that water is safe to drink, looking after animals and plants in the school grounds, and so on.
- Discuss ways in which human activities can destroy environments. Listen to the learners' responses and make sure that you mention dropping litter, wasting water, electricity and paper. Perhaps discuss also more environmentally friendly ways of travelling to school, for example walking or cycling or catching the school bus, rather than arriving by car.

Main activities

- Explain that some of the things humans do can have a positive impact and so are good for the environment. Other things that humans do can have a negative impact, which means that these things are not good for the environment.

- As part of the natural world, along with other animals and plants, we are inter-dependent – we are dependent on each other. Discuss how we need clean air and water and good growing conditions for plants, which animals eat, and ultimately they may form part of our diet (if we are meat-eaters).
- Give out photocopiable page 141 for the learners to record some of the negative and positive impacts that human activity can have on the environment.

Plenary

- Discuss the learners' responses to photocopiable page 141.
- Make sure that they understand what the environment is.

Success criteria

Ask the learners:

- What does 'environment' mean?
- Name one positive effect humans can have on the environment.
- Describe a negative effect that humans can have on the environment.
- What do we as humans need to survive?
- Why is it important for us to take good care of animals and plants?
- What things that are good or bad for the environment are happening locally?

Ideas for differentiation

Support: Discuss answers as a small group with these learners before they complete photocopiable page 141.

Extension: Ask these learners to identify something that is happening locally that is having either a positive or negative effect on the locality.

Name: _____

How do humans affect the environment?

Sometimes, human activity is good for the environment and helps to keep it a clean and safe place to be.

At other times, human activity can damage and spoil (or destroy) the environment.

Draw or write different ideas of positive and negative effects that humans can have on the environment.

Positive effect	Negative effect
1.	1.
2.	2.
3.	3.

Doing an environmental audit

- Explore how humans have positive and negative effects on the environment, e.g. loss of species, protection of habitats. (6Be1)

- Explore a number of ways of caring for the environment, e.g. recycling, reducing waste, reducing energy consumption, not littering, encouraging others to care for the environment. (6Be2)

- Consider how scientists have combined evidence from observation and measurement with creative thinking to suggest new ideas and explanations for phenomena. (6Ep1)

Resources

Photocopiable page 143.

Starter

- Ask the learners to discuss with talk partners: *How do we, in school, contribute to global environmental concerns?* Discuss the learners' responses and encourage them to elaborate on their answers.

- Have a class discussion to consider: *How can we reduce the negative impact we have in our school environment?* Again, listen to the learners' responses and expect them to give detailed explanations.

Main activities

- Explain that the purpose of this lesson is for the learners to conduct an environmental audit around the school. This will involve them working in groups or pairs to collect data about different areas. By doing this, they will work in a similar way to research scientists, collecting data, analysing results and suggesting new ideas and possible solutions to problems.

- Give out photocopiable page 143 and explain how to complete the table. Split the learners into pairs or groups and allocate the area you would like them to investigate. Alternatively, you may prefer to give each group a different starting point, but ask all groups to complete the table and then bring their findings back to class in a given time limit, for example 20 minutes.

- Discuss how the learners should behave as they move around the school, so as not to disturb other classes in lessons and adults working. Move around the school premises and site between groups while they carry out the audit, or work alongside the learners who need support in a small group.

- At a given time, meet back together in the classroom or at an agreed assembly point to share findings. Invite pairs or groups to share and comment on what they have found.

Plenary

- Ask for suggestions from each group on how to measure and compare consumption of water / electricity / fuel over time.

- Consider more environmentally friendly ways for the learners to travel to school (if possible).

- Think about ways of reducing the amount of paper used.

- Arrange a meeting with the headteacher to discuss the learners' concerns and ideas.

- Discuss ways in which scientists are suggesting we should be more environmentally aware.

Success criteria

Ask the learners:

- In which area do we appear to be most wasteful around school?
- In which area are we doing well?
- How could we improve in any area?
- How can we measure improvements over time?
- Are there any areas we need to raise awareness of around school?

Ideas for differentiation

Support: Work with these learners in a small group for this activity.

Extension: Ask these learners to write a letter to the headteacher expressing their concerns and making suggestions for improvements.

Name: _____

Environmental audit

1. Complete the table to show your estimate of how much waste there is in school. Tick (✓) to show how much waste you estimate there to be using the code below:

> 1 = unable to tell
>
> 2 = a lot of waste
>
> 3 = quite a lot of waste
>
> 4 = a little waste
>
> 5 = no waste

	1	2	3	4	5
Dripping taps					
Recycling paper					
Turning lights off					
Having heating on at sensible times					
Sorting waste, for example glass, plastic, food					
Walking or cycling to school					

2. What one thing do you think you should try to do better in school from your findings?

Collecting environmental data around school

Learning objectives

- Explore how humans have positive and negative effects on the environment, e.g. loss of species, protection of habitats. (6Be1)
- Explore a number of ways of caring for the environment, e.g. recycling, reducing waste, reducing energy consumption, not littering, encouraging others to care for the environment. (6Be2)
- Make a variety of relevant observations and measurements using simple apparatus correctly. (6Eo1)

Resources

Photocopiable page 143 (completed); photocopiable pages 145 and 146; weighing scales; black refuse bags; disposable rubber gloves; litter pickers; containers for collecting drops of water; measuring vessels – cylinders or beakers; internet access; publishing package (ICT).

Starter

- Invite the headteacher to class to respond to the letter written by the learners who did the extension activity in the previous lesson, or read out the headteacher's written response. Lead a discussion on how to provide evidence of current wastage, for example dripping taps and amount of litter.
- Thank the headteacher and assure them that the mission for your class will be to produce some more detailed data for consideration.

Main activities

- Give back or ask the learners to get out their filled-in photocopiable page 143 from the previous lesson. Ask them to discuss with talk partners ways in which any of these things can be easily measured to collect data.

- The learners will probably suggest weighing the amount of waste paper or collecting a volume of water from a dripping tap or collecting and sorting litter from around the school. On the basis of their suggestions (and their feasibility), explain that they will plan and measure the amount of waste in a particular area, then try to raise awareness around school of ways to reduce the amount of waste produced.
- Discuss possible data collection methods for each suggestion, but allow the learners to make the final decision about what they will measure and how. Organise the learners into pairs or small groups to carry out the activity.
- Give out photocopiable pages 145 and 146 for them to plan what they are going to do and to record their findings. Check their plans before allowing them to carry out the activity. Check and discuss their data once it is collected.

Plenary

- Invite different pairs or groups of learners to present their method and findings to the rest of the class. Allow the other learners to ask questions about what they have done already and what they intend to do next.
- Agree a timescale for collecting data.

Success criteria

Ask the learners:

- Which idea did you choose to investigate?
- What did you measure and how?
- What do you plan to do to raise awareness of this issue around school?
- Do you think that you can make a difference?

Ideas for differentiation

Support: Work with these learners in a small group. Select the idea to investigate, for example dripping taps.

Extension: Ask these learners to begin to produce a leaflet or web page to include all the information the class collects.

Name: _____

Collecting environmental data 1

1. Which idea will you investigate?

   Circle the idea below:

   | amount of litter   dripping taps   heating empty rooms   lights left on |
   | recycling (paper / glass / plastic / food)   ways of travelling to school |

2. How will you collect the data for this?

3. Record your data.

Name: _____

Collecting environmental data 2

1. Use your data to make a presentation to tell the rest of the class what you have found out. Think about the best way to do this – poster / leaflet / graph, and so on.

   Write your ideas here.

   ```
   _____

   _____

   _____

   _____

   _____

   _____

   _____

   _____

   _____

   _____
   ```

2. Use your data to make suggestions about how to improve the situation.

   Things that we could all do to try to make a difference:

   - _____

   - _____

   - _____

   - _____

   - _____

3. Arrange to collect more data after an agreed period of time and use it to compare it with this data.

 Cambridge Primary: Ready to Go Lessons for Science Stage 6 © Hodder & Stoughton Ltd 2013

Acid rain

Learning objectives

- Explore how humans have positive and negative effects on the environment, e.g. loss of species, protection of habitats. (6Be1)
- Explore a number of ways of caring for the environment, e.g. recycling, reducing waste, reducing energy consumption, not littering, encouraging others to care for the environment. (6Be2)
- Make a variety of relevant observations and measurements using simple apparatus correctly. (6Eo1)

Resources

Pictures (from books or the internet) of bacteria, a volcano, a power station burning a fossil fuel, vehicle exhaust fumes; 50 ml rainwater in a beaker; pre-prepared red cabbage indicator; 50 ml lemon juice or white vinegar in a beaker; large screw-top glass jar; distilled water; safety matches; sticks of chalk; white vinegar; glass beakers; photocopiable pages 148 and 149; pins.

Starter

- Show the learners the pictures one by one and discuss what types of pollution these all produce. (Bacteria and volcanoes produce nasty-smelling gases; fossil fuels produce soot and smoke and give off gases into the air; and vehicle exhaust emissions also release gases into the air.)
- Show the collected rainwater.

Main activities

- Explain that the smoke and / or fumes produced by bacteria, volcanoes, power stations and vehicles dissolve in rain as it falls, and make acid rain. Rain is made acid by carbon dioxide (the same gas that we all breathe out). Acid rain can affect trees as it falls, sometimes even causing them to die.
- Demonstrate the acidity of acid rain using red cabbage indicator. (This can be made by boiling red cabbage in a **small** amount of water – to make the indicator concentrated. Let it cool and store it – note that it does not keep well.)

- Give out photocopiable page 148 for the learners to complete as you do the demonstration. Mix a few drops of red cabbage indicator with the rainwater and note the colour.
- Mix a few drops of the indicator with the white vinegar or lemon juice. Note the colour changes.
- Now simulate acid rain: quarter-fill the glass jar with distilled water. Add a few drops of red cabbage indicator. Then light several safety matches in the glass jar above the water. When they finish burning, blow them out and quickly put the lid on the jar. Shake it and observe the colour change in the indicator in the jar.
- Then add some crushed chalk to the water, shake it again and observe what happens (chalk dissolves, some fizzing may be observed and more colour change). This demonstrates the effect of acid rain on soft rock.

Plenary

- Discuss the learners' observations and the colour changes observed.
- Talk about ways in which scientists today are suggesting we can reduce the amount of air pollution, which creates acid rain – using cleaner fuels, fewer vehicles, and so on.

Success criteria

Ask the learners:

- How does acid rain form?
- What kinds of pollution help create acid rain?
- Which gas contributes to acid rain being formed?
- What can acid rain do to buildings?

Ideas for differentiation

Support: Work closely with these learners in completing photocopiable page 148.

Extension: Give these learners photocopiable page 149. Alternatively use this as another class lesson on the effects of acid rain.

Name: _____

Using red cabbage indicator

1. Complete the table as your teacher does this demonstration.

You might need to leave some boxes blank.

Liquid	Colour	Colour with red cabbage indicator
red cabbage indicator		
rainwater		
lemon juice		
white vinegar		
acid rain (in jar)		

2. What happened when your teacher added crushed chalk to the jar?

_____

3. What colour does the red cabbage indicator turn in acid?

_____

4. Complete these sentences. Name a liquid each time.

a) _____ is a strong acid.

b) _____ is a weaker acid.

Cambridge Primary: Ready to Go Lessons for Science Stage 6 © Hodder & Stoughton Ltd 2013

Name: _____

What can acid rain do to buildings?

Some buildings are built of soft rocks, for example limestone. Chalk is an example of a limestone rock. This experiment will show you the effects of acid rain on soft rock.

You will need:

A pin, a stick of chalk, white vinegar, a glass beaker.

Method (what to do)

- Using the pin, carefully scratch a pattern or your initials on the stick of chalk.

- Pour some white vinegar into the glass beaker.

- Place the stick of chalk in the beaker.

- Observe what happens.

Results (what happened)

1. Choose the best way of recording your results.

2. What happens when acid rain falls on limestone buildings over time?

Collecting evidence of pollution in the air

Learning objectives

- Explore how humans have positive and negative effects on the environment, e.g. loss of species, protection of habitats. (6Be1)
- Explore a number of ways of caring for the environment, e.g. recycling, reducing waste, reducing energy consumption, not littering, encouraging others to care for the environment. (6Be2)
- Make a variety of relevant observations and measurements using simple apparatus correctly. (6Eo1)

Resources

Internet access or books; photocopiable pages 151 and 152; filter funnels and filter papers; beakers; water; plan of the school grounds; white ceramic tiles (or card, double-sided sticky tape and scissors); evergreen leaves.

Starter

- Show pictures (from the internet or books) that show sources of air pollution from around the world.
- Ask the learners to find with talk partners as many examples of sources of air pollution as they can from the pictures in two minutes.
- Discuss their responses. Perhaps compare living in a city to living the countryside: *Would there be any difference in the amount of pollution? Where would there be most pollution? Are there any big cities around the world that are badly polluted?*

Main activities

- Explain that pollution leaves dry, sooty materials in the air. In this lesson the learners will find out if there is any evidence of air pollution near to the school.
- Give out photocopiable pages 151 and 152. Go through these photocopiable pages with the learners, making sure that they know what to do. Alternatively, do each activity separately, or even over two lessons.

- Discuss what evergreen leaves are – these are leaves that remain on the tree, bush or plant all year round. Define the extent of the area in which the learners are allowed to look, or allocate different pairs or groups to particular areas of the school grounds. (Have ready some collected evergreen leaves in case the learners are unable to find any of their own.)
- Demonstrate (or ask the learners if they can remember) how to fold a filter paper for a funnel.
- Allocate (or allow the learners to choose) places to leave the white tiles or sticky-taped cards. You could mark these places on a plan of the school.
- Organise the learners into pairs or small groups to carry out the activity.

Plenary

- Bring the class back together and discuss their findings.

Success criteria

Ask the learners:

- What is an evergreen leaf?
- What was left behind when you filtered the water?
- What did you see on the white tile or sticky tape?
- Where did we find evidence of most pollution?
- What might be the biggest causes of air pollution in our neighbourhood?

Ideas for differentiation

Support: Either allow these learners to work in mixed-ability groups, or work with them in a small group.

Extension: Ask these learners to identify any evergreen leaves found by the learners.

Name: _____

Collecting evidence of pollution in the air 1

You will need:

Some evergreen leaves, water, a beaker, filter funnel and filter paper.

Method (what to do)

- Pick some evergreen leaves.

- Wash them in a beaker of water.

- Filter the water.

- Draw one of the leaves you collected in the box below.

- Name the leaf (if you can): _____

Results (what happened)

What was left after you filtered the water? Draw in the circle below.

Name: _____

Collecting evidence of pollution in the air 2

You will need:

A white tile (or a piece of card with tape on it, scissors, double-sided sticky tape – see method), pen.

Method (what to do)

- Write your name or initials on the white tile.

- If you do not have any white tiles, cut a strip of card and stick a piece of double-sided sticky tape to it along its length instead. Write your name or initials on the back of the card.

- Go to the place in the school grounds that your teacher tells you to go to, or the place that you chose.

Results (what happened)

1. We put our tile / card _____

2. Draw your results below.

3. Which place had most evidence of pollution? _____

Cambridge Primary: Ready to Go Lessons for Science Stage 6 © Hodder & Stoughton Ltd 2013

Environmental damage

Learning objectives

● Explore how humans have positive and negative effects on the environment, e.g. loss of species, protection of habitats. (6Be1)

● Explore a number of ways of caring for the environment, e.g. recycling, reducing waste, reducing energy consumption, not littering, encouraging others to care for the environment. (6Be2)

● Consider how scientists have combined evidence from observation and measurement with creative thinking to suggest new ideas and explanations for phenomena. (6Ep1)

Resources

A candle in a sand tray; a white tile or piece of safety glass; tongs; internet access or reference books; photocopiable pages 154 and 155.

Starter

• Look back at the results from the previous lesson. Ask the learners: *What was the main indicator of environmental pollution?* (Soot.)

• Ask the learners to discuss with talk partners where this soot came from. (Burning fuels, exhaust fumes, factory wastes.)

Main activities

• Demonstrate holding a white tile or piece of safety glass over a burning candle (using tongs). Make sure that the candle is securely fixed in a sand tray and that the learners are seated a safe distance away as you do the demonstration.

• Show the surface of the tile or glass blackened with soot.

• Ask the learners to think back to the unit of work on reversible and irreversible changes, where they observed a burning candle. Ask: *What else does a candle produce when it burns?* (Gases.) Explain that soot and gases are the waste products of burning.

• Look at pictures (on the internet or in books) of vehicle emissions in busy cities or from an aeroplane.

• Discuss if the learners know any ways in which such sources of pollution are dangerous to humans over time. (Exhaust fumes contain poisonous gases, which can cause allergies and intolerances in some people. Breathing in badly polluted air has been known to cause brain damage over time, particularly in children, who have smaller lungs. Treat this issue carefully as some of the learners may have family members with respiratory diseases. Also, learners at this age are very sensitive to thoughts of things that might kill them!)

• Explain that the use of unleaded fuels is being increased in many countries, which will reduce the level of poisonous gases in the air over time.

• Explain that in Ulaanbaatar, the capital city of Mongolia, new bus lanes have been introduced. Also, car drivers are limited to driving on no more than six days a week. Discuss the difference this might make. (Less traffic congestion, more people using public transport, less pollution from fewer vehicles, and so on.)

• Give out photocopiable page 154 or 155 for the learners to complete.

Plenary

• Discuss the learners' answers to the photocopiable pages.

• What might they suggest if they lived in Ulaanbaatar?

Success criteria

Ask the learners:

● What are the waste products of burning?
● Why is pollution a concern in big cities?
● How are some governments attempting to reduce levels of air pollution?
● How can pollution affect our health?

Ideas for differentiation

Support: Give these learners photocopiable page 154.

Extension: Ask these learners to write a leaflet encouraging the use of unleaded fuel.

Name: _____

Pollution

1. Complete this wordsearch. The words may read in any direction.

| breathe | burn | fumes | gases | petrol | soot |

s	o	b	r	k	y	m	p
l	i	r	u	g	r	t	o
k	p	e	t	r	o	l	y
e	r	a	p	t	n	o	g
c	u	t	g	l	i	b	a
i	v	h	r	e	s	p	s
i	s	e	m	u	f	l	e
n	e	p	y	t	o	o	s

2. Give one reason why pollution is bad for humans.

_____

_____

Cambridge Primary: Ready to Go Lessons for Science Stage 6 © Hodder & Stoughton Ltd 2013

Name: _____

Pollution in cities

1. Give three causes of pollution in big cities.

   a) _____

   b) _____

   c) _____

2. Write about some ways in which we are trying to reduce the amount of pollution around the world.

   _____

   _____

   _____

   _____

   _____

   _____

   _____

   _____

   _____

   _____

   _____

   _____

3. If you could ask your family to change one thing about creating pollution, what might it be?

   _____

   _____

   _____

Global warming (the greenhouse effect)

Learning objectives

- Explore how humans have positive and negative effects on the environment, e.g. loss of species, protection of habitats. (6Be1)
- Explore a number of ways of caring for the environment, e.g. recycling, reducing waste, reducing energy consumption, not littering, encouraging others to care for the environment. (6Be2)
- Consider how scientists have combined evidence from observation and measurement with creative thinking to suggest new ideas and explanations for phenomena. (6Ep1)

Resources

Internet access; globe; photocopiable page 157.

Starter

- Ask the learners to discuss with talk partners: *What is global warming? What are greenhouse gases?* This will give you an indication about how much or how little the learners already know.
- Listen to their responses and explain that in this lesson they will find out more about these things.

Main activities

- Go to http://news.bbc.co.uk/cbbcnews/hi/find_out/guides/world/global_warming/newsid_1575000/1575441.stm. This film clip gives good definitions of global warming and greenhouse gases.
- Give out photocopiable page 157 and explain the diagram of the Earth and its atmosphere. Explain how the Sun's rays penetrate the Earth's atmosphere and maintain the Earth's temperature, which ensures the survival of animals and plants.
- Explain (or ask the learners) about things that contribute to the Earth's temperature rising – carbon dioxide (CO_2) being exhaled, the Earth's growing population, so there is more CO_2 produced from people as well as from factories, vehicles, and so on. Describe the

CO_2 as forming a blanket that traps heat and means that over time (within the next century) scientists are predicting temperature rises of more than 6°C.

- Discuss what this will mean for us on Earth (use the globe as reference):
  - Glaciers and sea ice at the poles will melt. This means sea levels will rise, which will cause flooding in low-lying areas.
  - Lakes and rivers will dry up and in some areas this could create drought conditions. Ask: *How might this affect crop-growing in these areas?*
  - Winds could get stronger or travel from different directions, so there will be more tornadoes and hurricanes.
  - There will be less water available to us on Earth. Think about all the things we need water for on a daily basis.
  - Some animals and plants could become extinct due to the changes in their environment.

Plenary

- Go over the learners' responses to photocopiable page 157.

Success criteria

Ask the learners:

- What is global warming?
- Name some greenhouse gases.
- What might happen on Earth as a consequence of global warming?
- What are scientists suggesting as good alternatives to those things that contribute to global warming?

Ideas for differentiation

Support: Work with these learners in a small group to make sure that they understand the diagram.

Extension: Ask these learners to find out about some sources of alternative energy that are being trialled around the world, for example wind, wave, solar or hydro-electric power (HEP).

Name: _____

Global warming (the greenhouse effect)

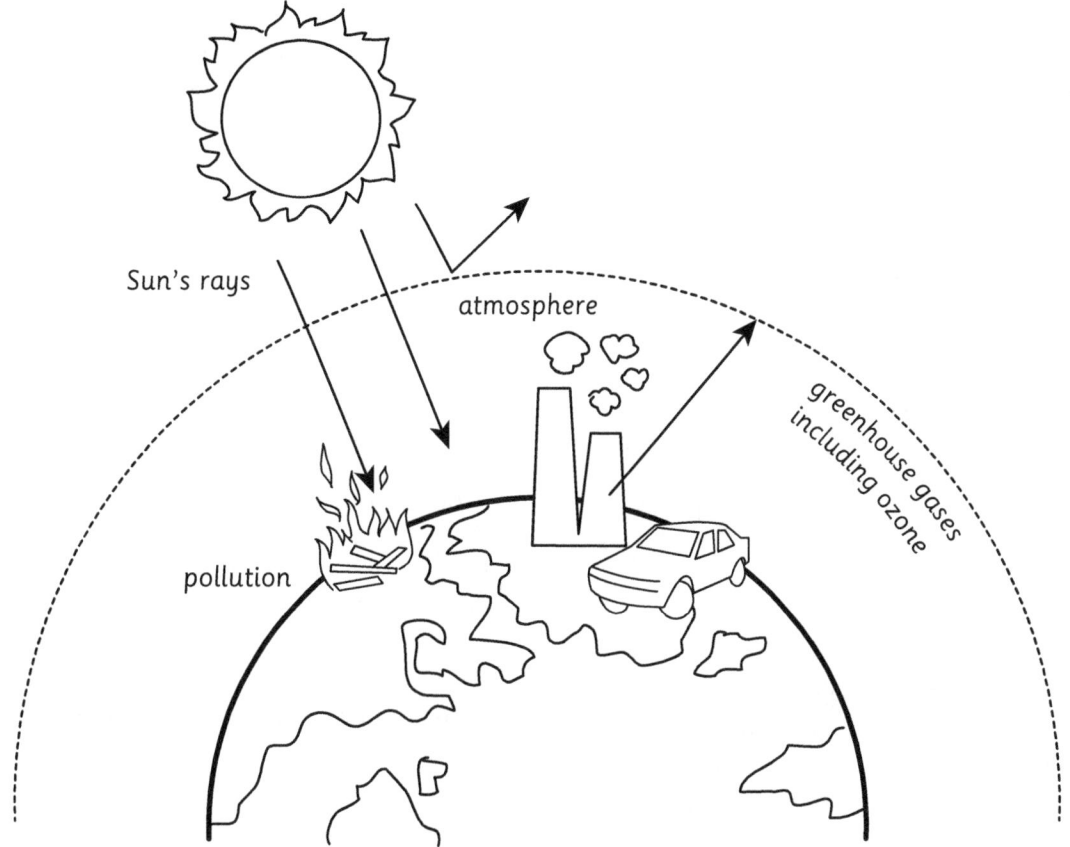

Complete the sentences to describe what happens during global warming.
Use the following words to help you.

atmosphere	deforestation	Earth	fossil fuels
greenhouse gases	pollution	Sun	

1. Global warming is how we describe the rise in the _____'s temperature.

2. The Earth is warmed by rays from the _____.

3. The Earth is surrounded by its _____.

4. The atmosphere contains _____.

5. Humans add to greenhouse gases by making _____.

6. Some pollution is made by burning _____.

7. In some parts of the world, forests are being destroyed.

    This is called _____.

The ozone layer

Learning objectives

● Explore how humans have positive and negative effects on the environment, e.g. loss of species, protection of habitats. (6Be1)

● Explore a number of ways of caring for the environment, e.g. recycling, reducing waste, reducing energy consumption, not littering, encouraging others to care for the environment. (6Be2)

● Consider how scientists have combined evidence from observation and measurement with creative thinking to suggest new ideas and explanations for phenomena. (6Ep1)

Resources

Globe; an aerosol spray; sticky labels; a pair of sunglasses; masks (the type that cover the eyes with eye-slits in them and elastic to hold them on); art materials (as requested); photocopiable page 159.

Starter

• Use the globe to remind the learners that the Earth is surrounded by its atmosphere and a layer of greenhouse gases. Introduce the word 'ozone' and describe it as the ozone layer, which protects the Earth from harmful UV (ultra-violet) rays from the Sun.

• Spray some of the contents from the aerosol can. Ask the learners how the aerosol works. (The gas inside the aerosol pushes out the liquid as a fine spray.)

Main activities

• Explain that scientists have discovered that holes are being formed in the ozone layer by the over-use of aerosols and some other chemicals called CFCs, which are used in refrigerators and freezers to keep them cool.

• Use this activity to describe how this happens. Choose some pairs of learners to be oxygen molecules. Give them each a sticker with the letter O on it. Choose another learner to be the Sun and give them the sunglasses to wear. Choose more learners to be CFCs and wear the masks.

• Each pair wearing the 'O' stickers is an oxygen molecule. The Sun learner must touch an oxygen pair on the shoulder, which makes them split apart and go off to find another oxygen molecule. When they join up as a three, they become an ozone molecule. This does not last long, as the oxygen atoms prefer to be in pairs, so the ozone molecule splits again. This happens all the time in the ozone layer.

• Then along come the nasty CFCs. They 'steal' oxygen atoms and make poisonous gases. (Invite the CFCs learners in to run around among the oxygen molecules.) This means that there is less ozone formed and it is not able to protect the Earth as well from the harmful UV rays from the Sun. Repeat the process as necessary.

• Organise the learners into small groups to perform this. Provide any props they reasonably request.

• Give out photocopiable page 159. Ask the learners to draw a cartoon of the process.

Plenary

• Watch the performances of the different groups.

• Talk about the United Nations Earth Summit. Every five years this is held for world leaders to discuss and report on initiatives to prevent global warming and on new methods to protect the environment.

Success criteria

Ask the learners:

● Where is the ozone layer?
● Why are CFCs harmful?
● What is ozone made from?
● How can we reduce damage in the ozone layer?

Ideas for differentiation

Support: Work with these learners in a small group or make them part of mixed-ability groups.

Extension: Ask these learners to act as narrators for other groups.

Name: _____

The ozone layer

1. Draw your own characters to match the labels below.

| The Sun | oxygen molecule | ozone | CFCs |

2. Draw a cartoon showing what happens in the ozone layer.

1.	2.	3.
4.	5.	6.

Deforestation

Starter

• Look at a world map or use a globe and identify areas where there are rainforests. Talk about what it is like inside a rainforest. Use a search engine to find pictures of a rainforest or use reference books.

• Ask the learners to work out with talk partners a definition of 'deforestation' (cutting down, clearing and removing forests).

• Ask the learners: *What are the causes of deforestation?* (They include farming, cattle ranching, logging, mining and extraction of oil or gas.)

Main activities

• Discuss each of the learners' suggestions in turn, for example agriculture is now thought to be responsible for a quarter of the Amazon rainforest clearance. When this happens, either for farming, mining or transport needs, fertile soil that is good for farming gets quickly washed away. Profits from large-scale farming and sales of the produce grown go back to big universal companies and don't benefit the locals who still live and work in or near the rainforest.

• Sometimes forests are cleared for new roads to be built. This separates areas within the rainforest – this could mean that some species of animals (monkeys, for example) might find it more difficult to get their food. If they don't eat as much food, they will not produce as much waste, which can contain seeds from plants that they have eaten, so fewer new plants will grow.

• When trees are cut down, animals and plants have to adapt to the different conditions – less shade, loss of habitat, and so on. Some may even become extinct as a result. Some trees take many years to grow.

• Give out photocopiable page 161 for the learners to complete about the effects of deforestation.

Plenary

• Discuss: *What are the consequences of deforestation?* (More animals and plants will become extinct, habitats will be destroyed, more carbon dioxide will be released into the atmosphere, and so on.)

Name: _____

Deforestation

1. Sometimes humans need to clear forests.
   For each reason, write what effect this has on the forest.

   a) Farming

   _____

   _____

   _____

   b) Road building

   _____

   _____

   _____

   c) Mining

   _____

   _____

   _____

2. What does the word 'extinct' mean?

   _____

   _____

3. Name an extinct animal.

   _____

What can we do?

Learning objectives

● Explore how humans have positive and negative effects on the environment, e.g. loss of species, protection of habitats. (6Be1)

● Explore a number of ways of caring for the environment, e.g. recycling, reducing waste, reducing energy consumption, not littering, encouraging others to care for the environment. (6Be2)

● Consider how scientists have combined evidence from observation and measurement with creative thinking to suggest new ideas and explanations for phenomena. (6Ep1)

Resources

Art materials or an ICT publishing package to make posters; photocopiable pages 163 and 164; felt-tipped pens or colouring pencils.

Starter

• Ask the learners to re-present their data from the environmental audit carried out at the beginning of this unit. Discuss any improvements and / or future recommendations for the school.

• Discuss ways in which the learners are aware of reducing, reusing and recycling things at home.

Main activities

• Talk about each of these areas (reducing, reusing and recycling) in turn.

• *What can we do to* **reduce** *our use of things that damage the environment?* (Buy a smaller car, try an electric car or a hybrid, take public transport, car-share, walk or cycle.) *Have there been any changes in the ways the learners travel to school?* This discussion may also lead to talking about reducing the use of chemicals or pesticides used by farmers and / or using less plastic and paper in packaging.

• Ask the learners: *What kinds of things can we easily* **reuse**? They will probably suggest clothes, vehicles, refrigerators, books, furniture, and so on. Some of them may know about places on

the internet where things can be sold instead of being thrown away. Suggest organising a 'swap shop' maybe for toys or books (with permission from home and the school management) to encourage other learners to begin to think about reusing things.

• Show the recycling logo. Discuss that this means that the object it appears on can be recycled. Discuss the recycling facilities (if any) available in school or locally. The main things to be recycled are usually plastic, glass, metals and paper. Ask: *How can we encourage better* **recycling** *at home and school?*

• Give out photocopiable pages 163 and 164 for the learners to record their own ideas about reducing, reusing and recycling on.

Plenary

• Reiterate the main messages that we need to look after the things we have and to be careful, not wasteful. We need to reduce, reuse and recycle.

Success criteria

Ask the learners:

● What things have improved in school?

● What should we do next?

● What are the three Rs of being environmentally aware?

● Tell me something that we should reduce our use of, how and why.

● What could we reuse more of?

● What things do we recycle?

Ideas for differentiation

Support: Work with these learners in a small group and ask them to give only one example of something you can reduce the use of, reuse and recycle.

Extension: Ask these learners to produce a poster to encourage being more environmentally friendly in school. Display these posters prominently around school.

Name: _____

Reduce, reuse, recycle

To be more environmentally friendly, we need to think about the following three things.

(**Reduce**)

1. Complete the table with things we should use less of. For each, give an alternative and a reason. The first one has been done for you.

Item	Alternative	Reason
pesticides	organic pesticides	stops poisoning soil

(**Reuse**)

2. Write an invitation to a friend to a swap shop! Use the 'Swap shop' page for this.

(**Recycle**)

3. Draw the recycling logo.

4. List some things you can recycle.

- _____

- _____

- _____

Name: _____

Swap shop

Write an invitation to a friend to bring some books or toys to your house that they could swap with things you don't want, instead of throwing them away.

Explain that this is to help avoid being wasteful.

Include a date and a time.

Remind them to ask permission to bring the things to swap.

Make it bright, beautiful and eye-catching!

INVITATION

Unit assessment

- Tell me one way in which we can look after our local environment.
- Describe a negative effect that human activity can have on the environment.
- What is the process called when forests are cut down?

- Name some materials that can be recycled.
- How can we reduce our use of fossil fuels?
- Which thing do you think would make the most difference in school if we did something more to care for the environment?

Summative assessment activities

Observe the learners while they participate in these activities. You will quickly be able to identify those who appear to be confident and those who may need additional support.

Recycling

This activity assesses the learners' understanding of recyclable materials.

You will need:

A rubbish bag or waste-bin (make sure this is clean), a selection of empty packets, tins, jars and boxes – typical of 'rubbish' from a bin at home (ensure it is all safe and that there are no sharp edges), a digital camera.

What to do

- Ask the learners as individuals to sort through the rubbish and re-group it for recycling.
- Discuss with them their choices as they carry out the activity or when they have finished.
- Differentiate the activity by asking learners of different abilities to sort more or fewer items.
- Take a digital photograph for evidence, if required.

Making compost

This activity assesses the learners' understanding of using recyclable materials to create compost.

You will need:

Plastic or paper cups; compost; magnifying glass; plastic gloves; soil; twigs; leaves; newspaper; water.

What to do

- Show the learners the compost. Ask them if they know what it is. If they don't, tell them and explain that this is nature's way of recycling! Allow them to look at it using a magnifying glass.
- Ask them to use the other things provided to make their own compost. Talk to them about each ingredient and why they have included it.
- List their ingredients and make notes about what the learners say about them.

Distribute photocopiable page 166. The learners should work independently, or with the usual adult support they receive in class.

Name: _____

Being environmentally friendly

1. Tick (✓) the ways in which you can be more environmentally friendly.

    a) Eat more sweets.

    b) Recycle paper, cans, glass and plastic.

    c) Buy recycled paper products.

    d) Stay up later at night.

    e) Encourage your parents to drive hybrid or electric cars.

    f) Walk or cycle to school.

    g) Save electricity by turning off electrical items when you are not using them.

    h) Clean your teeth and leave the tap running.

    i) Take showers instead of baths.

    j) Always use both sides of the paper.

2. Write down one way in which you already reduce or reuse or recycle things.

Cambridge Primary: Ready to Go Lessons for Science Stage 6 © Hodder & Stoughton Ltd 2013

Measuring mass

Learning objectives

- Distinguish between mass measured in kilograms (kg) and weight measured in newtons, noting that kilograms are used in everyday life. (6Pf1)
- Recognise and use units of force, mass and weight and identify the direction in which forces act. (6Pf2)
- Use tables, bar charts and line graphs to present results. (6Eo3)

Resources

A variety of different weighing scales – digital and analogue (kitchen scales, bathroom scales, spring balances); a variety of objects to weigh (classroom objects, food items, and so on); graph paper and markers or interactive whiteboard and software package to draw graphs; photocopiable pages 168 and 169.

Starter

- Ask the learners to discuss with talk partners the things we weigh in everyday life. Suggestions might include babies, our own weight, food at the supermarket.
- Talk about the different kinds of weighing machines we can use to find the weight of the things the learners suggest.
- If possible, have available some different weighing machines for the learners to use, for example kitchen scales and bathroom scales – familiar sets that may be like those used at home.

Main activities

- Explain that although we say that we are weighing an object or person, what we are really measuring is what scientists call its mass. The word 'weight' is just an everyday term that we use. The units of mass are grams (g) and kilograms (kg): 1000 g = 1 kg.
- Ask for volunteers to measure their weight using the bathroom scales. Be sensitive about this, as some learners will be self-conscious about their weight. Never force learners to do something that might make them feel uncomfortable.

- Use the data collected to compile and draw a class bar chart showing the range of weights among the learners in the class. Do this as a paper exercise or on the computer, as resources allow.
- Explain that this gives the learners experience in weighing objects in grams and kilograms. Practise by demonstrating the weighing of several objects and recording the weight in grams only, then converting the weight into grams and kilograms, for example 1750 g = 1 kg 750 g.
- Give out photocopiable page 168 to the learners who need support. Give photocopiable page 169 to all the other learners.

Plenary

- Invite the learners to share some of their measurements with the rest of the class. Discuss any disagreements and re-weigh if necessary to agree the true weight.

Success criteria

Ask the learners:

- What does the symbol 'g' stand for?
- What is the symbol for kilograms? (kg)
- What is the scientific word that we should use instead of saying 'weight'? (Mass.)
- Which was the largest mass recorded in class today?
- Which was the smallest mass measured in class today?

Ideas for differentiation

Support: Give these learners photocopiable page 168 and work with or provide adult support to work with this group.

Extension: Challenge these learners to find and weigh several items at home and record their masses in grams and kilograms.

Name: _____

Measuring mass

1. Complete the table to show the weight in grams (g)
    of several objects from around the classroom.

    Complete the column 'Weight in g'.

    Your teacher will help you to complete the final column.

Object	Weight in g	Weight in kg and g

2. Now complete these sentences.

    a) The heaviest object I weighed was _____.

    b) The _____ was the lightest object.

        It weighed _____.

    c) How many grams are there in 1 kg? _____ g

3. If you would like to, weigh yourself on the bathroom scales.

    My weight in kg is _____ kg.

Cambridge Primary: Ready to Go Lessons for Science Stage 6 © Hodder & Stoughton Ltd 2013

Name: _____

Measuring mass

1. Complete the table to show the weight in kilograms (kg) and grams (g) of several objects from around the classroom.

Object	Weight in kg and g

2. Now answer these questions.

   a) Which object had the greatest mass? _____

   b) Which object weighed the least? _____

3. If you would like to, weigh yourself on the bathroom scales.

   My weight in kg is _____ kg.

4. Complete this equation:

   _____ g = _____ kg

The story of gravity

Learning objectives

- Distinguish between mass measured in kilograms (kg) and weight measured in newtons, noting that kilograms are used in everyday life. (6Pf1)
- Recognise and use units of force, mass and weight and identify the direction in which forces act. (6Pf2)
- Make a variety of relevant observations and measurements using simple apparatus correctly. (6Eo1)

Resources

Flipchart and markers or whiteboard; bathroom scales that measure in newtons; newton meters (forcemeters); photocopiable pages 171 and 172.

Starter

- Ask the learners to recall from the work on famous scientists in Stage 5 anything that they can about Sir Isaac Newton. Create a mind map on the flipchart or whiteboard of their recollections and any additional information they suggest.

Main activities

- Remind the learners that the correct scientific word to use is 'mass' when we weigh something. It is a common, accepted mistake in everyday life to use the term 'weight'. Mass is a measurement of how much matter an object contains. The scientific unit of weight is not the kilogram; it is the newton. Weight is a type of force, measured in newtons (N).

- Tell the story of Newton when an apple apparently fell on his head, which caused him to think about what made that happen and ask 'Why did the apple fall?' This, ultimately, led to his discovering the force of gravity. The weight of any object is the result of the pull of gravity on it.

- Explain to the learners that the force of gravity pulls on the mass of an object. The bigger the mass of the object, the bigger the force made by gravity on it. This means that it will have a bigger weight. Gravity pulls things down towards the Earth. Demonstrate the bathroom scales that measure in newtons.

- Explain to the learners that, if they wish to, they can measure their weight in newtons using these scales. Invite a volunteer to help you demonstrate this.

- Then give out the newton meters (forcemeters) to the learners. Allow them time to explore and measure the weight in newtons of objects from around the classroom. Circulate the room, making sure that the learners understand how to use and calibrate the newton meter (forcemeter) if necessary and can read and interpret the scale.

- Give out photocopiable page 171 to all the learners except those who need an extension activity; use photocopiable page 172 for those learners. Explain that these photocopiable pages are to record some measurements in newtons. Check the predictions on photocopiable page 172 before allowing the learners to make the measurements.

Plenary

- Discuss and share the measurements made by the learners using newton meters (forcemeters). Agree results and check their use of the equipment used.

Success criteria

Ask the learners:

- Who discovered gravity?
- What is the unit of force?
- How do we write newtons as a symbol?
- What effect does the force of gravity have on Earth?

Ideas for differentiation

Support: Help these learners by pre-selecting the objects to measure and the newton meter (forcemeter) to be used.

Extension: Provide these learners with copies of photocopiable page 172.

Name: _____

Measuring weight in newtons (N)

You will need:

A newton meter (forcemeter).

Method (what to do)

- Check that you know how to use the newton meter (forcemeter).

- Measure the weight in newtons of different objects from around the room.

- Complete the table to show your results.

Object	Weight in newtons (N)

- If you would like to, measure your weight in newtons on the newton scales.

   My weight is: _____ N or _____ kg (from the previous lesson).

Name: _____

Measuring weight in newtons (N)

You will need:

A newton meter (forcemeter).

Method (what to do)

- Check that you know how to use the newton meter (forcemeter).

- Predict the weight in newtons of different objects from around the room.

- Show your teacher your predictions.

- Measure the weight in newtons of the objects.

- Complete the table to show your results.

Object	Predicted weight in newtons (N)	Weight in newtons (N)

- From your results, can you work out how many grams approximately = 1 newton?

Show your working in the box below:

- If you would like to, measure your weight in newtons on the newton scales.

Compare this with your weight in kg from the previous lesson.

Weight in newtons = _____   Weight in kg = _____

Measuring in kilograms (kg) and newtons (N)

- Distinguish between mass measured in kilograms (kg) and weight measured in newtons, noting that kilograms are used in everyday life. (6Pf1)
- Recognise and use units of force, mass and weight and identify the direction in which forces act. (6Pf2)
- Make a variety of relevant observations and measurements using simple apparatus correctly. (6Eo1)

Resources

A variety of scales to measure in kg and N; newton meters (forcemeters); a variety of food items; plastic bags to hold items for measuring; photocopiable pages 174 and 175.

Starter

- Ask the learners to discuss with talk partners: *What is the difference between mass and weight?*
- Discuss the learners' responses and remind them that mass is what we measure when we say we weigh something in grams (g) and kilograms (kg), but that weight is a force, measured in newtons (N).

Main activities

- Explain that in this lesson the learners will be able to measure and compare mass in grams (g) and kilograms (kg) with weight measured in newtons (N).
- Ask some of the learners who did photocopiable page 172 what they found out.
- Show the learners a range of food items and objects to choose from. Make sure that the food packages have the mass in kilograms (kg) and grams (g) clearly shown, for example flour, biscuits, a can of food, a bag of sugar. Include other items, for example loose fruit and vegetables that do not have a clearly marked mass but will need to be measured.

- Either direct the learners to select (for example three packaged items and three non-packaged items), or pre-select for them, groups of objects and foodstuffs, and give each group of learners such a selection to measure. The selection will depend on the availability of equipment, resources and time.
- Give out photocopiable page 174 to the learners who need support and photocopiable page 175 to all the other learners. Check that they can remember how to use, read and calibrate the weighing devices available before they begin. Carry out the activities as outlined on the photocopiable pages.

Plenary

- Discuss the learners' findings. Find out if they have measured anything that weighs 1 newton (N).
- Use this information to conclude that 1 newton (N) = approximately 100 g.

Success criteria

Ask the learners:

- What units do we measure mass in?
- What are newtons a unit of?
- What was the largest weight you measured?
- What was the smallest mass you measured?
- How many grams (g) (approximately) does 1 newton (N) equal?

Ideas for differentiation

Support: Give these learners photocopiable page 174 and pre-select the items for them to measure. Help and supervise them as they make measurements.

Extension: Challenge these learners to find something that weighs exactly 1 newton (N) and 10 newtons (N) that they have not measured before.

Name: _____

Measuring in kilograms (kg) and newtons (N)

You will need:

A newton meter (forcemeter), kitchen scales, foods to measure, plastic bags – to put food in.

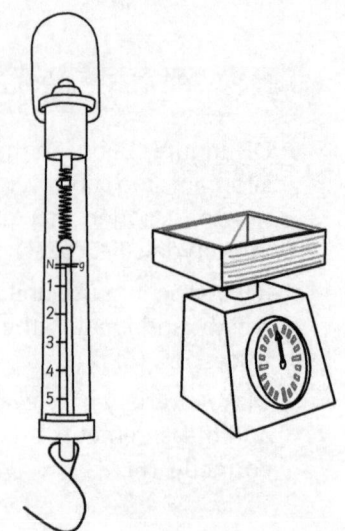

Method (what to do)

- Check that you know how to use the newton meter (forcemeter) and kitchen scales.

- Measure the foods in kilograms and in newtons.

Results (what happened)

- Complete the table to show your results.

Food	Mass in kg and g	Weight in newtons (N)

- Did you find anything that weighs 1 newton? What was it? _____

Cambridge Primary: Ready to Go Lessons for Science Stage 6 © Hodder & Stoughton Ltd 2013

Name: _____

Measuring in kilograms (kg) and newtons (N)

You will need:

Equipment for measuring mass and weight, food items, plastic bags to put food in.

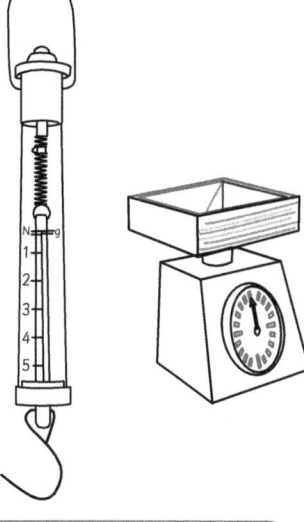

Method (what to do)

- Check that you know how to use the measuring equipment.
- Measure the mass and weight of the foods.
- Record your results.

Results (what happened)

- From your results, work out how much mass in g = 1 newton.

_____ g = 1 newton

Forces act in different directions

- Recognise and use units of force, mass and weight and identify the direction in which forces act. (6Pf2)
- Make a variety of relevant observations and measurements using simple apparatus correctly. (6Eo1)

Resources

Internet access; pictures (from the internet or books) of forces in action; photocopiable pages 177 and 178; colouring pencils.

Starter

- Ask the learners to discuss with talk partners: *Which direction does gravity act in? How do you know?*
- Remind the learners that in Stage 3 they learnt about forces being pushes and pulls.

Main activities

- Show some pictures of forces in action; these could include parachutes, something floating on water, a ball game, a cap on a bottle being unscrewed.
- For each picture, discuss which forces are in action and what happens when a force is applied. (A parachute falls to earth, a ship is designed to float, the ball helps to score in a sports game, the bottle will open, and so on.)
- Discuss that there may be more than one force acting on the object. Describe gravity as a pulling force. Ask: *Which pictures show objects being pulled to Earth?* Explain that gravity pulls things down to Earth and that this is why some things fall down or sink in water.
- Explain that the direction in which forces act is sometimes shown in diagrams by arrows – the bigger the force, the bigger the arrow. Demonstrate some examples of this, using the pictures available.

- Give out photocopiable pages 177 and 178 to the learners for them to draw on directional arrows of the main forces acting in the pictures. Give photocopiable page 177 to the learners who need support and photocopiable page 178 to all the other learners.

Plenary

- Discuss the learners' responses. Introduce the word **upthrust** and describe it as a push that opposes gravity. Explain that this is the reason that some things float on water – it is because the pull down of gravity is balanced by the upwardly acting force of upthrust.
- Ask the learners why they think that some objects sink in water, due to the effect of forces acting on them. Just listen to their ideas at this stage.
- Also introduce the phrase **air resistance**. Explain that this is the force that slows down a parachute falling to Earth. Gravity pulls it down to Earth, but air resistance is an upward force that pushes against the force of gravity.

Success criteria

Ask the learners:

- What kind of a force is gravity?
- Why do some things float?
- Which force balances gravity so that things float?
- What does air resistance do?

Ideas for differentiation

Support: Give these learners photocopiable page 177. Assist them by discussing the pictures before they draw the arrows.

Extension: Ask these learners to label as many different forces as possible in each picture, using different colours for different forces and writing a key to explain the colours.

The direction of forces

Forces can act in different directions.

All forces are either pushes or pulls.

In each picture, draw an arrow or some arrows to show the direction in which the forces are acting.

Name: _____

The direction of forces

Forces act in different directions.

All forces are either **pushes** or **pulls**.

1. In this picture, draw an arrow or some arrows to show the direction in which the forces are acting.

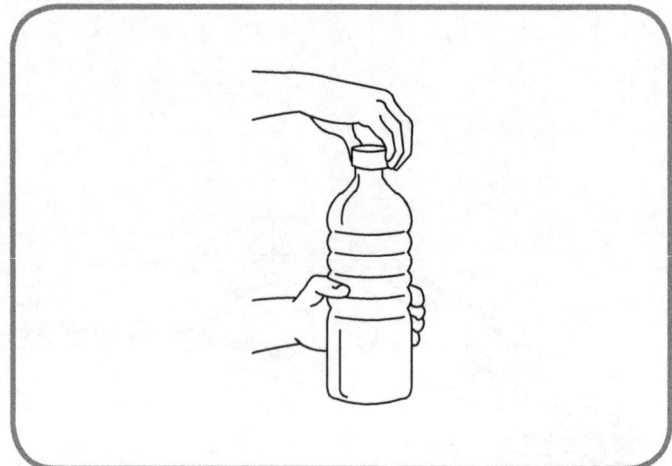

2. Describe how the force works to open the bottle.

    This is a different kind of push or pull.

    What kind of movement is it? _____

3. Now draw and label two pictures of your own that show forces in action with arrows.

Energy in movement

Learning objectives

● Understand the notion of energy in movement. (6Pf3)

● Make a variety of relevant observations and measurements using simple apparatus correctly. (6Eo1)

Resources

A large space; internet access; pictures of moving objects or people (from the internet or books); photocopiable pages 180 and 181.

Starter

• Take the class to a large space – the school hall or playground, for example. Direct the learners to perform lots of different moving activities – running, skipping, hopping, jumping, and so on. Ask them to get faster, slow down, change direction or stop.

• After each type of movement, talk about what they need to be able to move at all (food – and muscles). Discuss where they get their energy from (food) and that this is the fuel that their muscles use to help their bodies move.

• Discuss also what they feel like after a period of physical activity. They begin to feel tired, muscles ache, they need to take in food and / or water.

Main activities

• Return to the classroom and look at pictures (from the internet or books) of things, animals and people moving.

• Discuss each picture in turn, commenting on what is moving, where it gets its energy from and the type of movement that energy creates.

• Explain that energy is required to make something (human, animal or object) move. If there is no energy, there will be no movement. The fuel for our bodies is food. Other fuels power machines and make the whole machine or just a part of it move.

• Ask the learners to discuss with talk partners the different kinds of energy that make things move, for example electricity for motors, batteries for toys or gadgets.

• Give out photocopiable page 180 to the learners who need support. This photocopiable page contains pictures of things that move and the learners have to identify what is moving and where it gets its energy from.

• Give out photocopiable page 181 to all the other learners. They have to identify types of movement and sources of energy and suggest some of their own ideas.

Plenary

• Invite the learners to share some of their responses and their own ideas from completing the photocopiable pages. Correct any misconceptions or misunderstandings.

• Explain that energy can create movement. If there is no energy, there is no movement.

Success criteria

Ask the learners:

● Name a way in which the human body can move.

● Where do we get our energy from?

● How does this help us to move?

● What happens if there is no energy?

● Name some sources of energy for moving objects or machines.

Ideas for differentiation

Support: Give these learners photocopiable page 180. Discuss the pictures with them and agree the energy source before they complete the page.

Extension: Challenge these learners to find out what happens to make a car move in terms of energy.

Energy in movement

1. Look at this table showing things that move.

2. Complete the table.

Thing that moves	What is moving?	Where does it get its energy from?
an American footballer		
a helicopter		
a frog jumping		
riding a bicycle		

Cambridge Primary: Ready to Go Lessons for Science Stage 6 © Hodder & Stoughton Ltd 2013

Name: _____

Energy in movement

1. Look at this table showing things that move.

2. Complete the table.

3. Add one more idea of your own.

Thing that moves	What is moving?	What is its source of energy?
a fish swimming		
a runner		
an aeroplane		

Balanced and unbalanced forces

Starter

• Go to www.engineeringinteract.org/resources/parkworldplot/flash/concepts/balancedandun.htm. Demonstrate or ask the learners to volunteer to attempt each of the activities shown. Work through the activities, which demonstrate the effects of balanced and unbalanced forces affecting movement.

• Discuss each separate activity as you work through the animation.

Main activities

• Explain that when forces are balanced, there is no movement. The object, machine, animal or person is still and motionless. When something begins to move, speed up, slow down or change direction, this is possible because the forces are now unbalanced.

• Demonstrate by standing on a chair or on some stairs and dropping a ball of modelling clay. Ask the learners which forces are acting on the modelling clay as you hold it up in the air. (Gravity is pulling it down, but air resistance is pushing it up.) Drop the ball and ask the learners to explain what is different in terms of forces when you dropped it. (The force of gravity is greater than the force of air resistance, so the ball of modelling clay falls to the floor.)

• Discuss or demonstrate, if possible, different examples of unbalanced forces, for example a sinking boat. Load a toy boat with masses until it sinks. When it sinks it means that the force of gravity is greater than the upthrust, so the boat sinks.

• Explain that the reason a rocket can take off is that the thrust from its engines is greater than the force of gravity pulling it back to Earth.

• Go to www.huffingtonpost.co.uk and search for 'nasa rocket'. Watch the film clip of Nasa's latest rocket take off. Alternatively, look for the most recent rocket launches on the NASA website: www.nasa.gov.

• Give out photocopiable page 183 and explain to the learners that they have to say whether the forces are balanced or unbalanced, name the forces involved and say which is the strongest.

Plenary

• Discuss the learners' responses to photocopiable page 183.

• Make sure that the learners understand that forces always work in pairs. Talk about balanced forces (which means that something is still) and unbalanced forces (which generate movement).

Name: _____

Balanced and unbalanced forces

Complete the table.

Object	Balanced or unbalanced forces?	Forces involved	Which force is strongest?
a sinking ship			
a falling rock			
a parachute			
a table			
a hot air balloon floating in the air			

Friction

- Recognise friction (including air resistance) as a force which can affect the speed at which objects move and which sometimes stops things moving. (6Pf4)
- Choose what evidence to collect to investigate a question, ensuring that the evidence is sufficient. (6Ep5)
- Decide when observations and measurements need to be checked by repeating to give more reliable data. (6Eo2)

Resources

Newton meters (forcemeters); bricks or shoeboxes and masses; string; scissors; flipchart and markers; sticky notes; different surfaces; photocopiable page 185.

Starter

- Show the learners a heavy brick or a shoebox filled with masses. Tie the string around the brick or box and attach the newton meter (forcemeter) to it.
- Ask the learners to predict with talk partners the force in newtons that will be exerted if the brick (or box) is pulled along the surface of the table or desk. Record their predictions on sticky notes or a flipchart.
- Ask a learner to come forward and try it. Take the reading on the newton meter (forcemeter) and discuss how close their predictions were.

Main activities

- Remind the learners that friction is a force that acts between two surfaces (here we are using the brick / box and the table top). Compare the two different surfaces. Ask: *Are they rough or smooth? Does this make a difference?* Ask the learners to recognise and name different surfaces around the room and around school. (They will probably identify the floor [wood / carpet / matting], corridor and playground or playing field.)

- Explain that in this lesson they will be able to plan and carry out an investigation to find out which surface has the least friction. Tell them that they will have to choose which surfaces they want to test. Discuss factors and fair testing – remind them that only one factor should be changed each time (but don't tell them that this is the surface).
- Organise the learners into mixed-ability groups and give out photocopiable page 185 for their planning and recording.
- Check their method of recording results before they carry out the test. Circulate the room, helping, questioning and providing suggestions to make sure that the learners are successful.

Plenary

- Ask different groups of learners to share their conclusions. Do they agree?
- Discuss and decide on the best surface for reducing friction. (Compare measurements.)
- Explain that friction is a force that can make moving objects slow down or stop – it reduces movement. It happens when two surfaces rub against each other.

Success criteria

Ask the learners:

- Which surfaces did you test?
- Which surface had the least friction?
- Which surface had the most friction?
- What is friction?
- How can we reduce friction?

Ideas for differentiation

Support: Organise these learners into mixed-ability groups for this activity.

Extension: Ask these learners to find the best surface material for an entrance to school.

Name: _____

Investigating friction

1. Measure the force when a heavy object is pulled over different surfaces.

2. Which heavy object did you use?

    We used a _____ .

3. Show your results below.

Results (what happened)

4. Which was the best surface for reducing friction and why?

    _____

    _____

    _____

Air resistance

Starter

- Give each learner a piece of A4 paper. Ask them to drop it and observe how it falls.
- Now repeat the activity with the paper folded into four. Compare and discuss the difference.
- Ask the learners to screw the paper up into a ball and drop it again. Again, discuss and compare what happens.
- The learners should realise that changing the shape of the paper affects how it falls.
- Make a display of the different shapes of paper and ask the learners to write labels to describe the differences in how they fall through the air.

Main activities

- Explain that another force, rather like friction, is air resistance. Air resistance is friction between air and something else.
- Use a toy aeroplane to demonstrate the forces operating around an aircraft when it is airborne. Use sticky labels or arrows to label the toy as thrust going forwards as it flies through the air and air resistance pushing from the front to back of the aeroplane.
- Talk about streamlining and how altering the shape of aeroplanes and cars can create less air resistance and enable them to travel faster.

- Explain that you are going to have a competition to make paper aeroplanes to see which travels the furthest. Allow the learners to work in pairs or small groups to research, make and test their paper aeroplanes. Give them internet access or provide books and pictures with information about making paper aeroplanes.
- Within an agreed time limit, have them make their paper aeroplanes.
- Give out photocopiable page 187 for the learners to record what they do.

Plenary

- Hold a 'flying competition'. Discuss why the winning aeroplane was the winner – look at the shape and how streamlined it is.
- Remind the learners that streamlining reduces air resistance, so it can 'cut through' the air more easily.

Name: _____

Air resistance

1. Draw the different ways in which you dropped a piece of A4 paper.

a)	b)	c)

2. What difference did this make to how it fell through the air?

_____

_____

_____

3. Draw your paper aeroplane in the box below (or take a photograph).

4. How could you improve this model?

_____

_____

Making parachutes

● Recognise friction (including air resistance) as a force which can affect the speed at which objects move and which sometimes stops things moving. (6Pf4)

● Evaluate repeated results. (6Eo5)

● Use results to draw conclusions and to make further predictions. (6Eo7)

An outdoor space; large pieces of paper (flipchart paper); string; scissors; modelling clay; squares of paper; timers; photocopiable pages 189 and 190.

Starter

• Take the learners outside. Give each learner a piece of flipchart paper. Ask them to try to walk into the wind, or simply just to walk around, holding the paper in front of them.

• Discuss what it feels like. Explain that the force that they can feel pushing against them is air resistance.

Main activities

• Return to the classroom. Hand out the string, scissors, modelling clay and squares of paper to the learners in pairs or small groups. Give oral instructions for how to make a paper parachute. Give clear instructions on how to attach string to the corners of the canopy and how to attach the mass of modelling clay. Demonstrate how to do this as you give the instructions so that the learners can copy what you are doing.

• Try out several of these parachutes and time how long it takes for them to reach the ground from the time they were released into the air. Ask the learners to predict if they should take more time, less time or the same amount of time to fall from the same height. (They should take the same amount of time if they are identical.) Ask some of the learners to time the fall and the other learners to record the times.

• Discuss the results and any differences. Use this to highlight the importance of fair testing and that human error can sometimes have an effect. Ask: *Did you follow the instructions precisely?* Think about repeating results to get an average time to obtain more reliable results.

• Ask the learners to predict how the size of a parachute will affect the rate of fall.

• Give out photocopiable pages 189 and 190 for the learners to record their investigation. Organise them into mixed-ability groups for this activity. Carry out the activity and complete the photocopiable pages.

Plenary

• Invite groups in turn to present their findings to the rest of the class.

• Compare similarities and differences in their results.

• Predict the rate of fall of another parachute twice the size of the largest one used. If there is time, try this out.

Ask the learners:

● Which force acts downwards for a parachute?

● What is the opposing, upwards force?

● Which size of parachute fell the fastest?

● Which parachute took the longest time to land?

● Why did this happen?

● How does the size of the parachute affect the rate of fall?

Support: Either organise these learners into mixed-ability groups for the parachute investigation or work together with them in a small group.

Extension: Ask these learners to further investigate if the parachute material makes any difference to the speed of falling.

Name: _____

Making parachutes 1

square canopy

strings

knot
cover knot with
modelling clay

> Does the size of the canopy affect the rate of fall?

Prediction

I think that the _____ the canopy, the _____ the parachute will fall.

Diagram

Draw a picture of your parachute. Label the size of the canopy.

Method (what you did)

1. _____

2. _____

3. _____

4. _____

5. _____

6. _____

Name: _____

Making parachutes 2

square canopy

strings

knot
cover knot with
modelling clay

> Does the size of the canopy affect the rate of fall?

Results (what happened)

Conclusion (what you found out)

How does the size of the canopy affect the rate of fall?

The _____ the canopy, the _____ the rate of fall.

Explain why this is.

_____

_____

_____

Cambridge Primary: Ready to Go Lessons for Science Stage 6 © Hodder & Stoughton Ltd 2013

Unit assessment

- What is mass measured in?
- What units do we use to measure weight?
- How can we show on a diagram which direction a force is acting in?

- When something is still, are the forces acting on it balanced or unbalanced?
- What is friction?
- How is air resistance like friction?

Summative assessment activities

Observe the learners while they participate in these activities. You will quickly be able to identify those who appear to be confident and those who may need additional support.

Measuring mass and weight

This activity assesses the learners' ability to measure and record the mass and weight of an object.

You will need:

Different newton meters (forcemeters); a range of weighing scales; a selection of objects for measuring (different from those used in lessons previously).

What to do

- Ask the learners, working independently, to choose an object and record its mass and weight.
- Look for the correct use of units.
- Ask the learners to find an object that has a mass equal to a weight of 2 newtons.
- Then ask them to find an object with a mass equal to a weight of 1 newton.
- Differentiate this activity by asking easier or more difficult questions (measurements) depending on each learner's ability.

Moving vehicles

This activity assesses the learners' understanding and knowledge of forces that affect movement.

You will need:

A selection of toy vehicles; flipchart or large pieces of paper; markers.

What to do

- Ask the learners, working independently, to choose a vehicle. Let them place it on a large piece of paper and label the forces acting on it when it is still.
- Then ask them to make the vehicle move, speed up, slow down and change direction.
- Each time, they should place the vehicle on a different part of the paper and record the forces acting on it.
- Make notes regarding their understanding and knowledge of forces that affect movement.

Distribute photocopiable page 192. The learners should work independently, or with the usual adult support they receive in class.

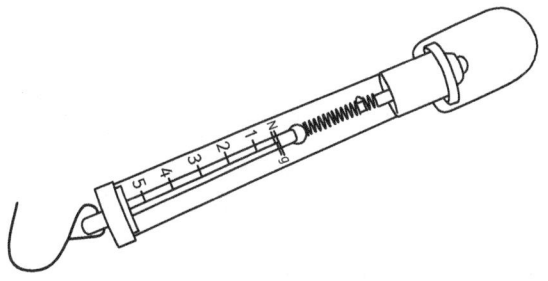

Name: _____

Parachutes

1. Label the forces acting on this parachute as it falls through the air.

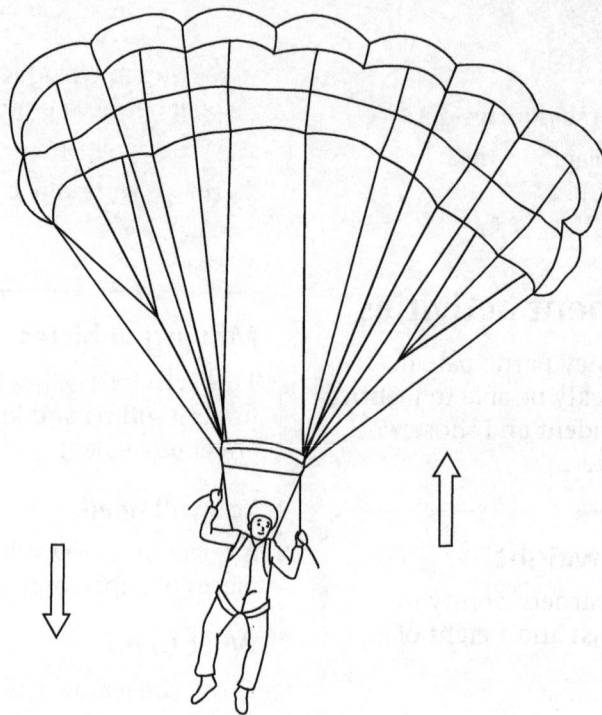

2. Class 6 made parachutes out of different materials.

    ● The parachutes were all the same size.

    ● They used the same mass on each one.

    ● They dropped them from the same height.

    ● They timed how long they took to fall.

    Here are their results.

Material	Time taken to fall in seconds
thin paper	20
thin cotton fabric	18
aluminium foil	15

Which material was best and why?

_____

_____

_____

Cambridge Primary: Ready to Go Lessons for Science Stage 6 © Hodder & Stoughton Ltd 2013